Streaming Data
Understanding the real-time pip

Streaming Data

UNDERSTANDING THE REAL-TIME PIPELINE

ANDREW G. PSALTIS

MANNING

SHELTER ISLAND

For online information and ordering of this and other Manning books, please visit
www.manning.com. The publisher offers discounts on this book when ordered in quantity.
For more information, please contact

> Special Sales Department
> Manning Publications Co.
> 20 Baldwin Road
> PO Box 761
> Shelter Island, NY 11964
> Email: orders@manning.com

Manning Publications Co.
20 Baldwin Road
PO Box 761
Shelter Island, NY 11964

Development editor:	Karen Miller
Technical development editor:	Gregor Zurowski
Project editor:	Janet Vail
Copyeditor:	Corbin Collins
Proofreader:	Elizabeth Martin
Technical proofreader:	Al Krinker
Typesetter:	Dennis Dalinnik
Cover designer:	Marija Tudor

ISBN: 9781617292286
Printed in the United States of America
1 2 3 4 5 6 7 8 9 10 – EBM – 22 21 20 19 18 17

brief contents

contents

preface

For as long as I can remember, I have been fascinated with speed as it relates to computing and am always trying to find a way to do something faster. In the late 1990s, when I spent most of my time writing software in C++, my favorite keyword was __asm, which means "the following block of code is in assembly language," and I understood what was happening at the machine level. I worked on mobile software in the early 2000s and again the story was how could we sync data faster or make things run faster on the PalmPilots and Windows CE devices we were using? At the time we had huge (by that day's standards, anyway) medical databases (around 25–50 MB in size) that required external cards on a PalmPilot to store and several applications that needed to provide interactive speed when searching and browsing the data.

As data volumes started to grow in the industries I was working in, I found myself at the perfect intersection of large data sets and speed to business insight. The data was growing in volume and being generated at faster and faster speeds, and business wanted answers to questions in shorter and shorter timeframes from the time data was being generated. To me, it was the perfect marriage: large data and a need for speed. Around 2001 I began to work on marketing analytics and e-commerce applications, where data was continuously being updated and we needed to provide insight into it in near real time. In 2009 I started working at Webtrends, where my love for speed and delivering insight at speed really matured. At Webtrends, analytics was our core business, and the idea of real-time analytics was just starting to catch the interest of our customers. The first project I worked on aimed to deliver key metrics in a dashboard within five minutes of a clickstream event happening anywhere in the world. At the time, that was *pushing the envelope*.

In 2011 I was part of an emerging products team. Our mission was to continue to push the idea of real-time analytics and try to disrupt our industry. After spending time researching, prototyping, and thinking through our next step, a perfect storm occurred. We had been looking at Apache Kafka, and then in September 2011 Apache Storm was open sourced. We immediately started to run like crazy with it. By winter we had early-adopter customers looking at what we were building. At that point we never looked back and set our sights on delivering on a Service Level Agreement (SLA) that was, in essence: "From click to dashboard in three seconds or less, globally!" After many months and a lot of work by what became a much larger team, we delivered on our promise and won the Digital Analytics New Technology of the Year award (www.digitalanalyticsassociation.org/awards2013). I was deeply involved in building and architecting all aspects of this solution, from the data collection to the initial UI (which was affectionately called "Bare Bones," due to my lack of UI skills).

We continued our pursuit and began looking at Spark Streaming when it was still part of the Berkley AMPLab. Since those days I have continued to pursue building more and more streaming systems that deliver on the ultimate goal of delivering insights at the speed of thought. Today I continue to speak internationally on the topic and work with companies across the globe in designing, building, and solving streaming problems.

Even today I still see a widespread lack of understanding of all the pieces that go into building and delivering a streaming system. You can usually find references to pieces of the stack, but rarely do you find out how to think through the entire stack and understand each of the tiers.

It is therefore with great pleasure that I have tried in this book to share and distill this real-world experience and knowledge. My goal has been to provide a solid foundation from which you can build and explore a complete streaming system.

acknowledgments

First, I want to thank my family for their support during the writing of this book. There were many weekends and nights of "Sorry, I can't help with the garden (or play lacrosse or go to the get-together)—I need to write." I'm sure that wasn't easy for my children to hear; nor was it always easy for my wife to buffer and pick up my slack. Through all the highs and lows that go into this process their support never wavered and they remained a constant source of encouragement and inspiration. For this I owe a tremendous debt of gratitude to my wife and children; a simple thank you cannot express it enough.

Thanks to Karen, my development editor, for her endless patience, understanding, and willingness to always talk things through with me throughout this entire journey. To Robin, my acquisition editor, for believing in me, nurturing the idea of this book, and being a sounding board to make sure the train was staying on the tracks during some rough patches in the early days. To Bert, for his teachings on how to tell a story, how to find the right level of depth with a narrative, and pedagogical insight into the construction of a technical book. To my technical development editor Gregor, whose very thoughtful and insightful feedback helped craft this book into what it is today. Lastly, but certainly not least, thanks to the entire Manning team for the fantastic effort to finally get us to this point.

Thanks also to all the people who bought and read early versions of the manuscript through the MEAP early access program, to those who contributed to the Author Online forum, and to the countless reviewers for their invaluable feedback, including Andrew Gibson, Dr. Tobias Bürger, Jake McCrary, Rodrigo Abreu, Andy Keffalas,

John Guthrie, Kosmas Chatzimichalis, Giuliano Bertoti, Carlos Curotto, Andy Kirsch, Douglas Duncan, Jeff Smith, and Sergio Fernández González, Jaromir D.B. Nemec, Jose Samonte, Jan Nonnen, Romit Singhai, Chris Allan, Jonathan Thoms, Steven Jenkins, Lee Gilbert, Amandeep Khurana, Charlie Gaines. Without all of you, this book wouldn't be what it is today.

Many others contributed in various different ways. I can't mention everyone by name because the acknowledgments would just roll on and on, but a big thank you goes out to everyone else who had a hand in helping make this possible!

about this book

The world of real-time systems has been around for a long time; for many years real-time and/or streaming was solely the domain of hardware real-time systems. Those are systems where if an SLA isn't met, there is potential loss of life. Over the last decade near-real-time systems have emerged and grown at an amazing rate. Everywhere you look you can find examples of data streaming: social media, games, smart cities, smart meters, your new washing machine, and the list goes on. Consider the following: Today if a byte of data were a gallon of water, an average home would be filled within 10 seconds; by the year 2020, it will only take 2 seconds. Making sense of and using such a deluge of data means building streaming systems.

Focusing on the big ideas of streaming and real-time data, the goals of this book are two-fold: The first objective is to teach you how to think about the entire pipeline so you're equipped with the skills to not only build a streaming system but also understand the tradeoffs at every tier. Secondly, this book is meant to provide a solid launching point for you to delve deeper into each tier, as your business needs require or as your interest pulls you.

How to use this book

Although this book was designed to read from start to finish, each chapter provides enough information so that you can read and understood it on its own. Therefore if want to understand a particular tier, you should feel comfortable jumping straight to that chapter and then using what you learned there as your base for deeper exploration of the other chapters.

Who should read this book

This book is perfect for developers or architects and has been written to be easily accessible to technical managers and business decision makers—no prior experience with streaming or real-time data systems required. The only technical requirement this book makes is that you should feel comfortable reading Java. The source code is written in Java, as is the example code that accompanies chapter 9

Roadmap

The roadmap of this book is represented in figure 1. A synopsis of each chapter follows.

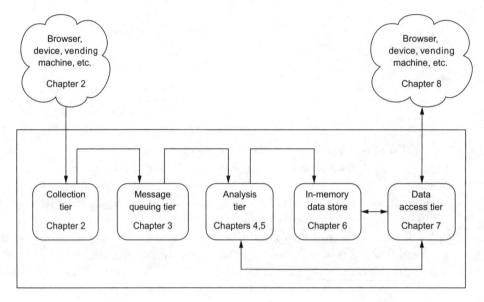

Figure 1 Architectural blueprint with chapter mappings

Chapter 1 introduces the architectural blueprint of the book, which tells you where we are in the pipeline and serves as a great map if you need to jump from tier to tier. After laying out this blueprint, chapter 1 defines a real-time system, explores the differences between real-time and in-the-moment systems, and briefly touches on the importance of security (which could be its own book).

Chapter 2 explores all aspects of collecting data for a streaming system, from the interaction patterns through scaling and fault-tolerance techniques. This chapter covers all the relevant aspects of the collection tier and prepares you to build a scalable and reliable tier.

Chapter 3 is all about how to decouple the data being collected from the data being analyzed by using a message queuing tier in the middle. You will learn why you need a message queuing tier, how to understand message durability and different message delivery semantics, and how to choose the right technology for your business problem.

Chapter 4 dives into the common architectural patterns of distributed stream-processing frameworks, covering topics such as what message delivery semantics mean for this tier, how state is commonly handled, and what fault tolerance is and why we need it.

Chapter 5 jumps from discussing architecture to querying a stream, the problems with time, and the four popular summarization techniques. If chapter 4 is the *what* for distributed stream-processing engines, chapter 5 is the *how*.

Chapter 6 discusses options for storing data in-memory during and post analysis. It doesn't spend much time discussing disk-based long-term storage solutions because they're often used out of band of a streaming analysis and don't offer the performance of the in-memory stores.

Chapter 7 is where we start to discuss what to do with the data we have collected and analyzed. It talks about communications patterns and protocols used for sending data to a streaming client. Along the way we'll find out how to match up our business requirements to the various protocols and how to choose the right one.

Chapter 8 explores concepts to keep in mind when building a streaming client. This is not a chapter on just building an HTML web app; it goes much deeper into lower-level things to consider when designing the client side of a streaming system.

Chapter 9 . . . at this point, if you have read all the way through, congrats! A lot of material is covered in the first eight chapters. Chapter 9 is where we make it all come to life. Here we build a complete streaming data pipeline and discuss taking our sample to production.

About the code

All the code shown in the final chapter of this book can be found in the sample source code that accompanies this book. You can download the sample code free of charge from the Manning website at www.manning.com/books/streaming-data. You may also find the code on GitHub at https://github.com/apsaltis/StreamingData-Book-Examples.

The sample code is structured as separate Maven projects, one for each of the tiers we walk through in chapter 9. Instructions for building and running the software are provided during the walkthrough in chapter 9.

All source code in listings or in the text is in a `fixed-width font` like this to separate it from ordinary text. In some listings, the code is annotated to point out the key concepts.

About the author

ANDREW PSALTIS is deeply entrenched in streaming systems and obsessed with delivering insight at the speed of thought. He spends most of his waking hours thinking about, writing about, and building streaming systems. He helps customers of all sizes build and/or fix complex streaming systems, speaks around the globe about streaming, and teaches others how to build streaming systems. When he's not busy being busy, he's spending time with his lovely wife, two kids, and watching as much lacrosse as possible.

Author Online

The purchase of *Streaming Data* includes free access to a private forum run by Manning Publications where you can make comments about the book, ask technical questions, and receive help from the author and other users. To access and subscribe to the forum, point your browser to www.manning.com/books/streaming-data. This page provides information on how to get on the forum once you're registered, what kind of help is available, and the rules of conduct in the forum.

Manning's commitment to our readers is to provide a venue where meaningful dialogue between individual readers and between readers and the author can take place. It's not a commitment to any specific amount of participation on the part of the author, whose contribution to the book's forum remains voluntary (and unpaid). We suggest you try asking him challenging questions, lest his interest stray!

The Author Online forum and the archives of previous discussions will be accessible from the publisher's website as long as the book is in print.

About the cover illustration

The figure on the cover of *Streaming Data* is captioned "Habit of a Moor of Morrocco in winter in 1695." The illustration is taken from Thomas Jefferys' *A Collection of the Dresses of Different Nations, Ancient and Modern* (four volumes), London, published between 1757 and 1772. The title page states that these are hand-colored copperplate engravings, heightened with gum arabic. Thomas Jefferys (1719–1771) was called "Geographer to King George III." He was an English cartographer who was the leading map supplier of his day. He engraved and printed maps for government and other official bodies and produced a wide range of commercial maps and atlases, especially of North America. His work as a mapmaker sparked an interest in local dress customs of the lands he surveyed and mapped, which are brilliantly displayed in this collection.

Fascination with faraway lands and travel for pleasure were relatively new phenomena in the late 18th century and collections such as this one were popular, introducing both the tourist as well as the armchair traveler to the inhabitants of other countries. The diversity of the drawings in Jefferys' volumes speaks vividly of the uniqueness and individuality of the world's nations some 200 years ago. Dress codes have changed since then and the diversity by region and country, so rich at the time, has faded away. It is now often hard to tell the inhabitant of one continent from another. Perhaps, trying to view it optimistically, we have traded a cultural and visual diversity for a more varied personal life. Or a more varied and interesting intellectual and technical life.

At a time when it is hard to tell one computer book from another, Manning celebrates the inventiveness and initiative of the computer business with book covers based on the rich diversity of regional life of two centuries ago, brought back to life by Jeffreys' pictures.

Part 1

A new holistic approach

Today data is streaming all around us, with new data sources coming online daily. If you're not yet faced with building a real-time data system, it's only a matter of time before you will be. More and more businesses will depend on being able to process and make decisions on streams of data. This first part of this book looks at a streaming system, from the point of ingestion all the way through delivering the data for display or consumption by other systems.

Chapter 1 begins by introducing streaming data and laying the foundation of terms we will use. *Streaming data* and *real time* may mean different things to different people. This chapter clarifies how *we* will use these terms and defines our architectural blueprint that we will use as our guide throughout the book. At the end of chapter 1, we glance at how the security relates to streaming systems.

The entry point to a streaming system is the collection or ingestion of data. The patterns of collecting data and preventing data loss are our focus throughout chapter 2.

Upon ingestion of data we need to move it as fast as we can to a message queue (or as some may call it, message *buffer*). The technology used in this tier comes with various levels of durability, delivery semantics, and impact on the producers and consumers of data. Chapter 3 looks at best practices and how to take these features into account.

Chapter 4 covers analyzing streaming data. The focus here is on in-flight data analysis, common stream-processing architectures, and the key features common to all distributed stream-processing engines.

When using a distributed stream-processing engine, there are numerous things you need to think about. In particular, time. How should you think about time with a streaming system? You'll find out in chapter 5. This chapter also discusses four powerful and common summarization techniques used when analyzing a stream of data.

After analyzing a stream of data, you may need to store it. That may sound strange—why store it, we're processing a stream of data! Chapter 6 discusses why you may need to store data, what you may want to store, and how to do it properly when processing a stream of data.

By the time we get to chapter 7 we have collected, queued, analyzed, and potentially stored the stream of data. Now the discussion moves on to the next tier of making this data available, because in the end we need to provide the results of our analysis to another system that can take action on the stream.

Chapter 8 wraps up part 1 with a discussion of the core principles to consider when building a streaming client, introduces the web client, and closes out with a discussion of querying a stream of data—something your users are going to want.

Introducing
streaming data

1

This chapter covers

- Differences between real-time and streaming data systems
- Why streaming data is important
- The architectural blueprint
- Security for streaming data systems

Data is flowing everywhere around us, through phones, credit cards, sensor-equipped buildings, vending machines, thermostats, trains, buses, planes, posts to social media, digital pictures and video—and the list goes on. In a May 2013 report, Scandinavian research center Sintef estimated that approximately 90% of the data that existed in the world at the time of the report had been created in the preceding two years. In April 2014, EMC, in partnership with IDC, released the seventh annual Digital Universe study (www.emc.com/about/news/press/2014/20140409-01.htm), which asserted that the digital universe is doubling in size every two years and would multiply 10-fold between 2013 and 2020, growing from 4.4 trillion gigabytes to 44 trillion gigabytes. I don't know about you, but I find those numbers hard to comprehend and relate to. A great way of putting that in perspective also comes from that report: today, if a byte of data were a gallon of water, in only 10 seconds there would be enough data to fill an average home. In 2020, it will only take 2 seconds.

Although the notion of Big Data has existed for a long time, we now have technology that can store all the data we collect and analyze it. This does not eliminate the need for using the data in the correct context, but it is now much easier to ask interesting questions of it, make better and faster business decisions, and provide services that allow consumers and businesses to leverage what is happening around them.

We live in a world that is operating more and more in the *now*—from social media, to retail stores tracking users as they walk through the aisles, to sensors reacting to changes in their environment. There is no shortage of examples of data being used today as it happens. What is missing, though, is a shared way to both talk about and design the systems that will enable not merely these current services but also the systems of the future.

This book lays down a common architectural blueprint for how to talk about and design the systems that will handle all the amazing questions yet to be asked of the data flowing all around us. Even if you've never built, designed, or even worked on a real-time or Big Data system, this book will serve as a great guide. In fact, this book focuses on the big ideas of streaming and real-time data. As such, no experience with streaming or real-time data systems is required, making this perfect for the developer or architect who wants to learn about these systems. It's also written to be accessible to technical managers and business decision makers.

To set the stage, this chapter introduces the concepts of streaming data systems, previews the architectural blueprint, and gets you set to explore in-depth each of the tiers as we progress. Before I go over the architectural blueprint used throughout the book, it's important that you gain an understanding of real-time and streaming systems that we can build upon.

1.1   What is a real-time system?

Real-time systems and *real-time computing* have been around for decades, but with the advent of the internet they have become very popular. Unfortunately, with this popularity has come ambiguity and debate. What constitutes a real-time system?

Real-time systems are classified as *hard*, *soft*, and *near*. The definitions I use in this book for *hard* and *soft real-time* are based on Hermann Kopetz's book *Real-Time Systems* (Springer, 2011). For *near real-time* I use the definition found in the Portland Pattern Repository's Wiki (http://c2.com/cgi/wiki?NearRealTime). For an example of the ambiguity that exists, you don't need to look much further than Dictionary.com's definition: "Denoting or relating to a data-processing system that is slightly slower than real-time." To help clear up the ambiguity, table 1.1 breaks out the common classifications of real-time systems along with the prominent characteristics by which they differ.

You can identify hard real-time systems fairly easily. They are almost always found in embedded systems and have very strict time requirements that, if missed, may result

Table 1.1 Classification of real-time systems

Classification	Examples	Latency measured in	Tolerance for delay
Hard	Pacemaker, anti-lock brakes	Microseconds–milliseconds	None—total system failure, potential loss of life
Soft	Airline reservation system, online stock quotes, VoIP (Skype)	Milliseconds–seconds	Low—no system failure, no life at risk
Near	Skype video, home automation	Seconds–minutes	High—no system failure, no life at risk

in total system failure. The design and implementation of hard real-time systems are well studied in the literature, but are outside the scope of this book. (If you are interested, check out the previously mentioned book by Hermann Kopetz.)

Determining whether a system is soft or near real-time is an interesting exercise, because the overlap in their definitions often results in confusion. Here are three examples:

- Someone you are following on Twitter posts a tweet, and moments later you see the tweet in your Twitter client.
- You are tracking flights around New York using the real-time Live Flight Tracking service from FlightAware (http://flightaware.com/live/airport/KJFK).
- You are using the NASDAQ Real Time Quotes application (www.nasdaq.com/quotes/real-time.aspx) to track your favorite stocks.

Although these systems are all quite different, figure 1.1 shows what they have in common.

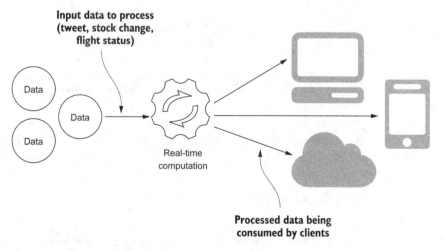

Figure 1.1 A generic real-time system with consumers

In each of the examples, is it reasonable to conclude that the time delay may only last for seconds, no life is at risk, and an occasional delay for minutes would not cause total system failure? If someone posts a tweet, and you see it almost immediately, is that soft or near real-time? What about watching live flight status or real-time stock quotes? Some of these can go either way: what if there were a delay in the data due to slow Wi-Fi at the coffee shop or on the plane? As you consider these examples, I think you will agree that the line differentiating soft and near real-time becomes blurry, at times disappears, is very subjective, and may often depend on the consumer of the data.

Now let's change our examples by taking the consumer out of the picture and focusing on the services at hand:

- A tweet is posted on Twitter.
- The Live Flight Tracking service from FlightAware is tracking flights.
- The NASDAQ Real Time Quotes application is tracking stock quotes.

Granted, we don't know how these systems work internally, but the essence of what we are asking is common to all of them. It can be stated as follows:

> Is the process of receiving data all the way to the point where it is ready for consumption a soft or near real-time process?

Graphically, this looks like figure 1.2.

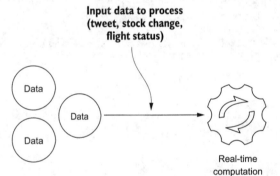

Real-time computation   **Figure 1.2   A generic real-time system with no consumers**

Does focusing on the data processing and taking the consumers of the data out of the picture change your answer? For example, how would you classify the following?

- A tweet posted to Twitter
- A tweet posted by someone whom you follow and your seeing it in your Twitter client

If you classified them differently, why? Was it due to the lag or perceived lag in seeing the tweet in your Twitter client? After a while, the line between whether a system is soft or near real-time becomes quite blurry. Often people settle on calling them real-time. In this book, I aim to provide a better way to identify these systems.

1.2 Differences between real-time and streaming systems

It should be apparent by now that a system may be labeled soft or near real-time based on the perceived delay experienced by consumers. We have seen, with simple examples, how the distinction between the types of real-time system can be hard to discern. This can become a larger problem in systems that involve more people in the conversation. Again, our goal here is to settle on a common language we can use to describe these systems. When you look at the big picture, we are trying to use one term to define two parts of a larger system. As illustrated in figure 1.3, the end result is that it breaks down, making it very difficult to communicate with others with these systems because we don't have a clear definition.

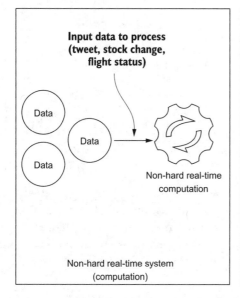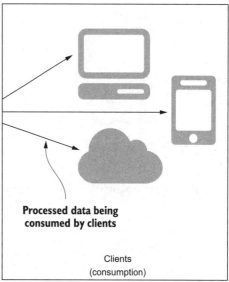

Figure 1.3   Real-time computation and consumption split apart

On the left-hand side of figure 1.3 we have the non-hard real-time service, or the *computation* part of the system, and on the right-hand side we have the clients, called the *consumption* side of the system.

> **DEFINITION: STREAMING DATA SYSTEM**   In many scenarios, the computation part of the system is operating in a non-hard real-time fashion, but the clients may not be consuming the data in real time due to network delays, application design, or a client application that isn't even running. Put another way, what we have is a non-hard real-time service with clients that consume data when they need it. This is called a *streaming data system*—a non-hard real-time

system that makes its data available at the moment a client application needs it. It's neither soft nor near—it is streaming.

Figure 1.4 shows the result of applying this definition to our example architecture from figure 1.3.

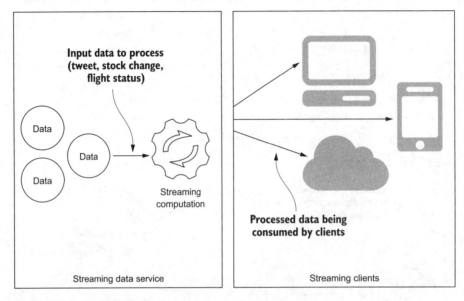

Figure 1.4   A first view of a streaming data system

The concept of streaming data eliminates the confusion of soft versus near and real-time versus not real-time, allowing us to concentrate on designing systems that deliver the information a client requests at the moment it is needed. Let's use our examples from before, but this time think about them from the standpoint of streaming. See if you can split each one up and recognize the streaming data service and streaming client.

- Someone you are following on Twitter posts a tweet, and moments later you see the tweet in your Twitter client.
- You are tracking flights around New York using the real-time Live Flight Tracking service from FlightAware.
- You are using the NASDAQ Real Time Quotes application to track your favorite stocks.

How did you do? Here is how I thought about them:

- *Twitter*—A streaming system that processes tweets and allows clients to request the latest tweets at the moment they are needed; some may be seconds old, and others may be hours old.
- *FlightAware*—A streaming system that processes the most recent flight status data and allows a client to request the latest data for particular airports or flights.

■ *NASDAQ Real Time Quotes*—A streaming system that processes the price quotes of all stocks and allows clients to request the latest quote for particular stocks.

Did you notice that doing this exercise allowed you to stop worrying about soft or near real-time? You got to think and focus on what and how a service makes its data available to clients at the moment they need it. Thinking about it this way, you can say that the system is an *in-the-moment* system—any system that delivers the data at the point in time when it is needed. Granted, we don't know how these systems work behind the scenes, but that's fine. Together we are going to learn to assemble systems that use open source technologies to consume, process, and present data streams.

1.3 The architectural blueprint

With an understanding of real-time and streaming systems in general under our belt, we can now turn our attention to the architectural blueprint we will use throughout this book. Throughout our journey we are going to follow an architectural blueprint that will enable us to talk about all streaming systems in a generic way—our pattern language. Figure 1.5 depicts the architecture we will follow. Take time to become familiar with it.

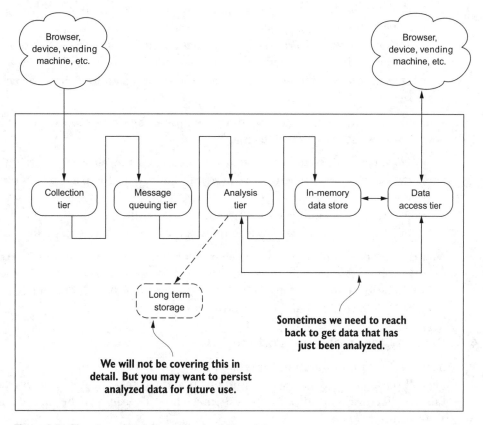

Figure 1.5　The streaming data architectural blueprint

As we progress, we will zoom in and focus on each of the tiers shown in figure 1.5 while also keeping the big picture in mind. Although our architecture calls out the different tiers, remember these tiers are not hard and rigid, as you may have seen in other architectures. We will call them tiers, but we will use them more like LEGO pieces, allowing us to design the correct solution for the problem at hand. Our tiers don't prescribe a deployment scenario. In fact, they are in many cases distributed across different physical locations.

Let's take our examples from before and walk through how Twitter's service maps to our architecture:

- *Collection tier*—When a user posts a tweet, it is collected by the Twitter services.
- *Message queuing tier* —Undoubtedly, Twitter runs data centers in locations across the globe, and conceivably the collection of a tweet doesn't happen in the same location as the analysis of the tweet.
- *Analysis tier*—Although I'm sure a lot of processing is done to those 140 characters, suffice it to say, at a minimum for our examples, Twitter needs to identify the followers of a tweet.
- *Long-term storage tier*—Even though we're not going to discuss this optional tier in depth in this book, you can probably guess that tweets going back in time imply that they're stored in a persistent data store.
- *In-memory data store tier*—The tweets that are mere seconds old are most likely held in an in-memory data store.
- *Data access*—All Twitter clients need to be connected to Twitter to access the service.

Walk yourself through the exercise of decomposing the other two examples and see how they fit our streaming architecture:

- *FlightAware*—A streaming system that processes the most recent flight status data and allows a client to request the latest data for particular airports or flights.
- *NASDAQ Real Time Quotes*—A streaming system that processes the price quotes of all stocks and allows clients to request the latest quote for particular stocks.

How did you do? Don't worry if this seemed foreign or hard to break down. You will see plenty more examples in the coming chapters. As we work through them together, we will delve deeper into each tier and discover ways that these LEGO pieces can be assembled to solve different business problems.

1.4   *Security for streaming systems*

As you reflect on our architectural blueprint, you may notice that it doesn't explicitly call out security. Security is important in many cases, but it can be overlaid on this architecture naturally. Figure 1.6 shows how security can be applied to this architecture.

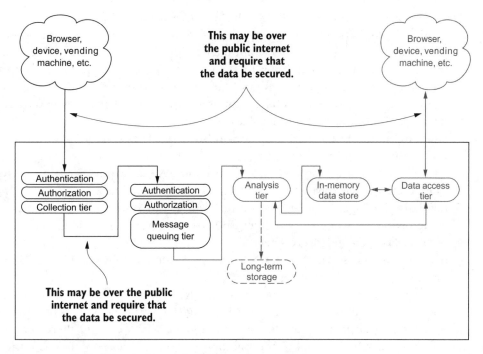

Figure 1.6   The architectural blueprint with security identified

We won't be spending time discussing security in depth, but along the way I will call it out so you can see how it fits and think about what it may mean for the problems you're solving. If you're interested in taking a deeper look at security and distributed systems, see Ross Anderson's *Security Engineering: A Guide to Building Dependable Distributed Systems* (Wiley, 2008). This book also is available for free at www.cl.cam.ac.uk/~rja14/book.html.

1.5   *How do we scale?*

From a high level, there are two common ways of scaling a service: vertically and horizontally.

Vertical scaling lets you increase the capacity of your existing hardware (physical or virtual) or software by adding resources. A restaurant is a good example of the limitations of vertical scaling. When you enter a restaurant, you may see a sign that tells you the maximum occupancy. As more patrons come in, more tables may be set up and more chairs added to accommodate the crowd—this is scaling vertically. But when the maximum capacity is reached, you can't seat any more customers. In the end, the capacity is limited by the size of the restaurant. In the computing world, adding more memory, CPUs, or hard drives to your server are examples of vertical scaling. But as with the restaurant, you're limited by the maximum capacity of the system, physical or virtual.

Horizontal scaling approaches the problem from a different angle. Instead of continuing to add resources to a server, you add servers. A highway is a good example of horizontal scaling. Imagine a two-lane highway that was originally constructed to handle 2,000 vehicles an hour. Over time more homes and commercial buildings are built along the highway, resulting in a load of 8,000 vehicles per hour. As you might imagine (and perhaps have experienced), the results are terrible traffic jams during rush hour and overall unpleasant commutes. To alleviate these issues, more lanes are added to the highway—now it is horizontally scaled and can handle the traffic. But it would be even more efficient if it could expand (add lanes) and contract (remove lanes) based on traffic demands. At an airport security checkpoint, when there are few travelers TSA closes down screening lines, and when the volume increases they open lines up. If you're hosting your service with one of the major cloud providers (Google, AWS, Microsoft Azure), you may be able to take advantage of this elasticity—a feature they often call *auto-scaling*. The basic idea is that as demand for your service increases, servers are automatically added, and as demand decreases, servers are removed.

In modern-day system design, our goal is to have horizontal scaling—but that doesn't mean that we won't use vertical scaling too. Vertical scaling is often employed to determine an ideal resource configuration for a service, and then the service is scaled out. But in this book, when the topic of scaling comes up, the focus will be on horizontal, not vertical scaling.

1.6   Summary

Now that you have an idea of the architectural blueprint, let's see where we have been:

- We defined a real-time system.
- We explored the differences between real-time and streaming (in-the-moment) systems.
- We developed an understanding of why streaming is important.
- We laid out an architectural blueprint.
- We discussed where security for streaming systems fits in.

Don't worry if some of this is slightly fuzzy at this point, or if teasing apart the different business problems and applying the blueprint seems overwhelming. I will walk through this slowly over many different examples in the coming chapters. By the end, these concepts will seem much more natural.

We are now ready to dive into each of the tiers to find out what they're composed of and how to apply them in the building of a streaming data system. Which tier should we tackle first? Take a look at a slightly modified version of our architectural blueprint in figure 1.7.

We're going to take on the tiers one at a time, starting from the left with the collection tier. Don't let the lack of emphasis on the message queuing tier in figure 1.7

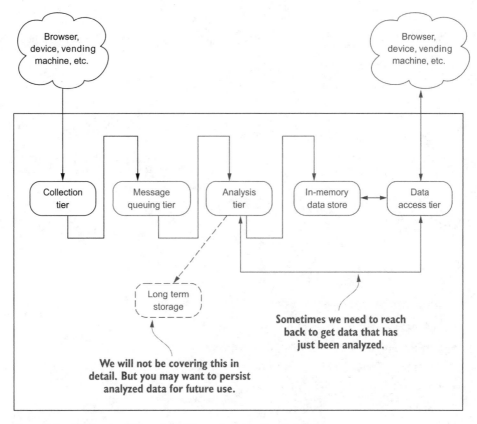

Figure 1.7 Architectural blueprint with emphasis on the first tier

bother you—in certain cases where it serves a collection role, I'll talk about it and clear up any confusion. Now, on to our first tier, the collection tier—our entry point for bringing data into our streaming, in-the-moment system.

Getting data from clients: data ingestion

On to our first tier: the *collection tier* is our entry point for bringing data into our streaming system. Figure 2.1 shows a slightly modified version of our blueprint, with focus on the collection tier.

This tier is where data comes into the system and starts its journey; from here it will progress through the rest of the system. In the coming chapters we'll follow the flow of data through each of the tiers. Your goal for this chapter is to learn about the collection tier. When you finish this chapter you will know about the collection patterns, how to scale, and how to improve the dependability of the tier via the application of fault-tolerance techniques.

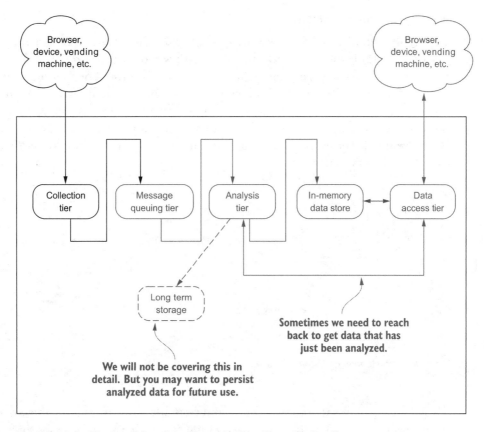

Figure 2.1 Architectural blueprint with emphasis on the collection tier

2.1 Common interaction patterns

Regardless of the protocol used by a client to send data to the collection tier—or in certain cases the collection tier reaching out and pulling in the data—a limited number of interaction patterns are in use today. Even considering the protocols driving the emergence of the Internet of Everything, the interaction patterns fall into one of the following categories:

- Request/response pattern
- Publish/subscribe pattern
- One-way pattern
- Request/acknowledge pattern
- Stream pattern

Let's discuss how you might collect data using each of these patterns.

2.1.1 *Request/response pattern*

This is the simplest pattern. It's used when the client must have an immediate response or wants the service to complete a task without delay. Every day you experience this pattern while browsing the web, searching for information online, and using your mobile device. Here's how the pattern works. First, a client makes a request to a service—this may be to take an action (such as send a text message, apply for a job, or buy an airline ticket) or to request data (such as perform a search on Google or find the current weather in their city). Second, the service sends a response to the client. Figure 2.2 illustrates this pattern.

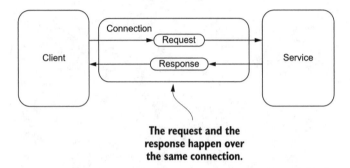

The request and the response happen over the same connection.

Figure 2.2   Basic request/response pattern

The simplicity of a synchronous request and response pattern comes at the cost of the client having to wait for the response and the service having to respond in a timely fashion. With modern-day services, this cost often results in an unacceptable experience for users. Imagine browsing to your favorite news or social site, and your browser tries to request all the resources in a synchronous fashion. Outside of basic services such as requesting the current weather, the potential delay is no longer tolerable. In many cases this can translate into lost revenue for merchants because users don't want to wait for the response.

Three common strategies can overcome this limitation: one on the client side, one on the service side, and one a combination of the two. Let's consider the client side first. A common strategy often taken by the client is to make the requests asynchronously; this approach is illustrated in figure 2.3.

With this adaptation the client makes the request of the service and then continues on with other processing while the service is processing the request. This is the pattern used by all modern web browsers: the browser makes many asynchronous requests for resources and renders the image and/or content as it arrives. This type of processing allows the client to maximize the time normally spent waiting on the response. The end result is an overall increase in the work performed by the client over a period of time. Implementing this type of pattern is relatively easy today because all modern programming languages and many of the frameworks you may

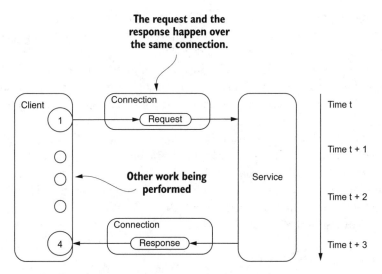

Figure 2.3 Client making asynchronous request to the service

use natively support performing the request asynchronously. This pattern is often called *half-async* because one half of the request response is done asynchronously.

Implementing this type of processing on the service end is also very common and is illustrated in figure 2.4.

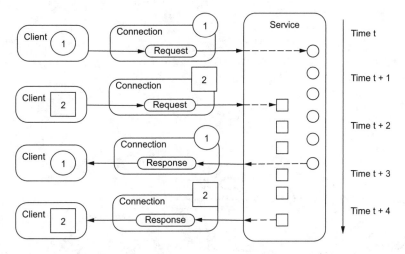

Figure 2.4 Service async request/response pattern

With the service-side half-async pattern, the service receives a request from a client, delegates the work to be done, and when the work is finished responds to the client. This type of processing results in the development of more scalable services, which can handle requests from many more clients. This type is also very common in all server-side development frameworks found today for all popular programming languages.

The last variation of this pattern, called *full-async*, occurs when both client and server perform their work asynchronously; the resulting flow is the same as that shown in figure 2.4. Today many modern clients and services operate in this full-async fashion.

Now let's walk through an example of using this pattern in the design and development of a streaming system. Imagine we work in the transportation industry, and last week while enjoying coffee with our friend Eric, who works in the automotive industry, we came up with an idea to provide a real-time traffic and routing service for all vehicles on the road. Our company would build the service, and Eric's company would build the streaming system that would reside inside the vehicles. We then sketched out what this solution would look like. Starting with the vehicle part of it, figure 2.5 shows our high-level drawing of the vehicle side of things.

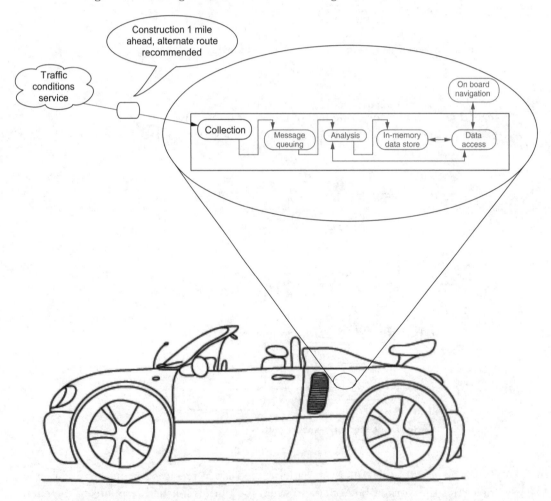

Figure 2.5   Receiving the response to the traffic conditions request with an on-board streaming system

For the vehicle, Eric is going to build an embedded streaming system to not only handle interacting with your traffic and navigation service but also to have the ability to interact with other services and perhaps vehicles.

The request/response pattern would work well for this scenario; in particular we'd want to choose the full-async variant. That way, our traffic and navigation service would be better positioned to handle requests from a lot of cars on the road at a single time. For Eric's on-board streaming system, the ability to asynchronously request data and process it as it arrives would be essential. By following this pattern, the streaming system would not be blocked waiting for a response from our service and could handle other data analysis pertinent to the vehicle.

At this point we're ready for Eric's team to build the vehicle side of things, and we're ready to build the traffic and routing service. If you're interested in learning more about this pattern, a good place to start is with Robert Daigneau's *Service Design Patterns* (Addison-Wesley, 2011).

2.1.2 *Request/acknowledge pattern*

There are times when you need to use an interaction pattern with similar semantics to the request/response pattern, but you don't need a response from the service. Instead, what you need is an acknowledgment that your request was received. The request/acknowledge pattern fits that need. Often the data sent back in the acknowledgment can be used to make subsequent requests, perhaps to check the status of the initial request or get a final response.

Imagine we're working with the marketing department for our company to make sure that on our e-commerce site we provide the right offer to the right person at the right time, with the goal of increasing their likelihood of making a purchase during their current visit. After further discussions with the marketing team, we settled on a solution that would constantly update a visitor's propensity-to-buy score during their visit. With this dynamic score available, our site can make the right offer at any time to influence their decision to purchase. Figure 2.6 shows how this looks from a high level.

Let's walk through the flow of data illustrated in figure 2.6. As the visitor is browsing our site, we're collecting data about each page they visit and every link they click. The unique thing we're doing that's particular to the request/acknowledge pattern occurs on the first page they visit. On this page our collection tier returns an acknowledgment that can be used in future requests. Unlike the request/response pattern, which may return as response success or failure, the request/acknowledge pattern returns data that can be used in future requests. In this case the acknowledgment is nothing more than a unique identifier, but it plays an important role. The acknowledgment can be used on all subsequent pages the visitor visits. When we call the propensity service, we can pass the unique identifier we obtained on the very first visit. With the unique identifier, which identifies the visitor, our propensity service can determine and return the visitor's current propensity-to-purchase score.

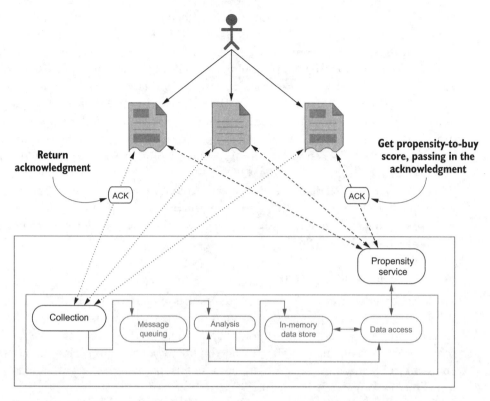

Figure 2.6   Visitor browsing while data is collected and their propensity-to-buy score is updated

I realize I'm leaving out a lot of the details of how we got from collection to a propensity score, but it will become clearer in the coming chapters. The key takeaway is that with the request/acknowledge pattern, a client makes a request of a service, asking for an action to be taken, and in turn receives an acknowledgment token that can be used in future requests. We experience this pattern every day in real life. For example, when you purchase an item online, you're often given a confirmation number, which you can then use to check on the status of your order.

If you're interested in learning more about this pattern, see Gregor Hohpe and Bobby Woolf's *Enterprise Integration Patterns* (Addison-Wesley, 2003).

2.1.3   Publish/subscribe pattern

This is a common pattern with message-based data systems; the general flow is shown in figure 2.7.

The general data flow as illustrated in figure 2.7 starts with a producer publishing a message to a broker. The messages are often sent to a *topic*, which you can think of as a logical grouping for messages. Next, the message is sent to all the consumers

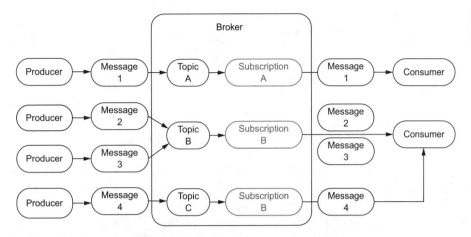

Figure 2.7 General data flow for the publish/subscribe message pattern

subscribing to that topic. There's a subtlety in this last step, covered in-depth in chapter 3. For now, know that some technologies follow the data flow as illustrated here, pushing messages to consumers. But with other technologies, the consumer pulls messages from the brokers. It may not be obvious initially, but often a producer publishing a message doesn't mean that it needs to subscribe to a topic. Nor is it required that a subscriber produce a message.

Let's walk through an example of how this protocol can be used and see its impact on our collection tier. After the success of our joint venture with Eric's company, we started to think about how we can take our in-vehicle traffic and routing service to the next level. After considering several ideas, we settled on the idea of making it social. In addition to the vehicle requesting traffic information and routing, it would send real-time traffic updates back to the service and subscribe to the real-time traffic reports from other vehicles traveling along the same route. Figure 2.8 shows the flow of messages we're talking about.

For simplicity, the figure shows only a handful of cars acting as producers and sending their current traffic conditions to the broker, and a single car acting as the consumer. If this were real, you could imagine how each producer would also consume and analyze all the data. By using the publish/subscribe pattern we're able to decouple the sender of the traffic data from its consumer. As we scale this simple example—four cars sending data and one consuming data—to all cars in the United States, you can imagine how important the decoupling this pattern provides is. If you're interested in learning some of the finer points about this pattern, a good place to start is *Enterprise Integration Patterns*, mentioned previously.

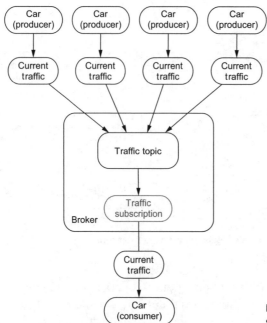

Figure 2.8 **Current traffic publish/ subscribe message pattern**

2.1.4   *One-way pattern*

This interaction pattern is commonly found in cases where the system making the request doesn't need a response. You may also see this pattern often referred to as the "fire and forget" message pattern. In some cases this pattern has distinct advantages and may be the only way for a client to communicate with a service. It's similar to the request/response and request/acknowledge patterns in the way a message is sent from the client to the service. The major difference is that the service doesn't send back a response. In the other patterns the client knows the request was received and processed; with the one-way pattern the client doesn't even know whether the request was received by the service.

You may be wondering how or where a pattern that has zero guarantees that the message was even received by a service can be useful. It's useful in environments where a client doesn't have the resources or the need to process a request. For example, think of the available data about the servers in your data center. You'd like for the server to send data about how much memory and CPU are being used every 10 seconds. You don't need the server to take any action or even worry about the result; it's purely producing data as fast as possible.

Examples of this interaction pattern appear all around us and will continue to grow along with the proliferation of the Internet of Everything, which is infiltrating many aspects of our life, sports being no exception. A recent partnership between the NFL and Zebra (www.zebra.com/us/en/nfl.html) resulted in players during

Thursday Night Football games being outfitted with quarter-sized radio frequency identifier (RFID) tags on their equipment. Each tag transmits data, such as the athlete's movement, distance, and speed, approximately 25 times a second to the 20 RFID receivers installed in the stadium. Within half a second, the data is analyzed and relayed to the TV broadcast trucks to be used by commentators. In this scenario, the RFID tag, the client, doesn't need and doesn't have the resources to process a response from the RFID receiver.

With the data being sent 25 times a second, if, during one second, five samples were lost and were not received by the RFID receiver, would the resulting analysis be impacted? No, it would not. That's another noteworthy characteristic of this pattern— it is appropriate for and often found in environments where losing some data is tolerable in exchange for simplicity, reduced resource utilization, and speed. To learn more about this pattern, see Nicolai M. Josuttis's *SOA in Practice* (O'Reilly, 2007).

2.1.5 *Stream pattern*

This interaction style is quite different than all the others discussed so far. With all the other patterns, a client makes a request to a service that may or may not return a response. The stream pattern flips things around, and the service becomes the client. A comparison of this to the other patterns you've seen is illustrated in figure 2.9.

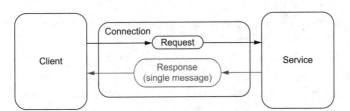

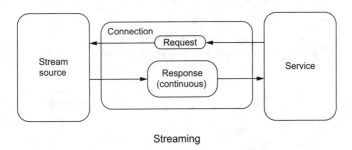

Figure 2.9 Comparing the request/response patterns to the stream pattern

There are a couple of important distinctions to point out when comparing the previous patterns (all the request/response optional patterns) with the stream pattern:

- With the request/response style of interaction as depicted at the top of figure 2.9, the client pushes data to the service in the request, and the service may respond. This response is grayed out in the diagram because the response is not required by some variations of this pattern. It boils down to a single request resulting in zero or one response. The stream pattern as depicted at the bottom of figure 2.9 is quite different; a single request results in no data or a continual flow of data as a response.

- In the request/response optional patterns a client external to the streaming system is pushing the message to it. In our previous examples this was a web browser, a car, or a phone—all clients that send a message to our collection tier. In the case of the stream pattern, our collection tier connects to a stream source and pulls data in. For example, you may be interested in building a streaming system to do sentiment analysis of tweets. To do so, you'd build a collection tier that establishes a connection to Twitter and consumes the stream of tweets.

This pattern is very interesting and powerful: ingesting a stream of data and producing another stream. With this you can quickly build a streaming analysis system that consumes publicly available data and in turn creates new streams of data based on your analysis. Unlike the other patterns, where you need to create or find clients to send a request to your service, with the stream pattern you can chose to connect to and process the data from a stream source.

Meetup.com provides an example input stream that you can use for exploring this interaction pattern or as input to a streaming system. This is also the stream we will use in chapter 9 when we build an end-to-end streaming system. The stream is composed of JSON events, each of which is generated every time someone RSVPs to a meetup. To see this input stream in action, open your favorite browser and go to http://stream.meetup.com/2/rsvps. In this case a simple, long-lived HTTP connection is established, and data is subsequently streamed back to your browser until you end the HTTP connection. In the data stream you'll see JSON events that are similar to the following listing.

See
code
listings
9.2–9.7

Listing 2.1 Example JSON stream event

```
{
    "venue": {
        "venue_name": "Chicago Symphony Center",          Venue-related
        "lon": -87.624402,                                 information
        "lat": 41.878904,
        "venue_id": 700306
    },
    "visibility": "public",
    "response": "yes",
    "member": {                           Member-related
        "member_id": 184005505,           information
```

```
                     "member_name": "Rifat"
        },
        "rsvp_id": 1631403261,
        "event": {
                "event_name": "Civic Orchestra Open Rehearsal w\/ Riccardo Muti -
        Brahm' s 4th Symphony",
                "event_id": "234044012",
                "time": 1474934400000
        },
        "group": {
                "group_topics": [{
                        "urlkey": "symphony",
                        "topic_name": "Symphony"
                }],
                "group_city": "Chicago",
                "group_country": "us",
                "group_id": 882009,
                "group_name": "Chicago Classical Music Events",
                "group_lon": -87.63,
                "group_urlname": "chicagosymphony",
                "group_state": "IL",
                "group_lat": 41.88
        }
}
```

Event-related information

Group-related information

Listing 2.1 is an example of using the stream interaction pattern. Imagine if you took this data and combined it with social data such as tweets about certain events or venues. Again, that's something that's hard to replicate with the other patterns. As you look at this data, I'm sure you'll come up with questions about it without combining it with other streams. Perhaps you want to count the top events people are RSVP'ing to or maybe the top groups by city. But let's not get ahead of ourselves—we're going to work through how to answer these and other questions in chapter 4. For now, it's enough to recognize this pattern and start to get a feel for this interaction pattern.[1]

2.2 Scaling the interaction patterns

Now that we've discussed each of the interaction patterns, let's see how we'd scale our collection tier and talk about some of the things to keep in mind when implementing it. We're going to keep the discussion at the level of the two categories we grouped the interaction patterns into before.

2.2.1 Request/response optional pattern

To discuss scaling this general pattern we'll continue with the example from our initial discussion of the request/response pattern, the real-time traffic and routing service for all vehicles on the road. To get a better sense for the scale of our idea of providing this service for all vehicles on the road in the United States, we'll consider the 2012 National Transportation Statistics report (the last complete year produced by

---

[1] If you're interested in learning more about this dataset, check out www.meetup.com/meetup_api/docs/stream/2/rsvps/#polling.

the Bureau of Transportation Statistics, www.rita.dot.gov). According to this report, approximately 253 million vehicles were registered in the United States and were driven approximately 2.966 trillion miles in 2012. This means that at any time during the almost 3 trillion miles driven by one of the 253 million vehicles, we may get a request for current traffic conditions and alternate route suggestions. At any moment we'll need to handle thousands and possibly millions of requests.

If you remember, chapter 1 talked about horizontal scaling being our overall goal for every tier of our streaming system. With this example and our use of the request/ response optional pattern, horizontal scaling will work very well for two reasons. First, with this pattern we don't have any state information about the client making the request, which means that a client can connect and send a request to any service instance we have running. Second—and this is a result of the stateless nature of this pattern—we can easily add new instances of this service without changing anything about the currently running instances. The mode of scaling stateless services is so popular that many cloud-hosting providers, such as Amazon, offer a feature called auto-scaling that will automatically increase or decrease the number of instances running, based on demand. On top of horizontal scaling, we also want our service to be *stateless*, which will allow any vehicle to make a request to any instance of our service at any time. This stateless trait is commonly found in systems that use this pattern. Taking horizontal scaling and statelessness into consideration, we arrive at figure 2.10, which shows these two aspects together.

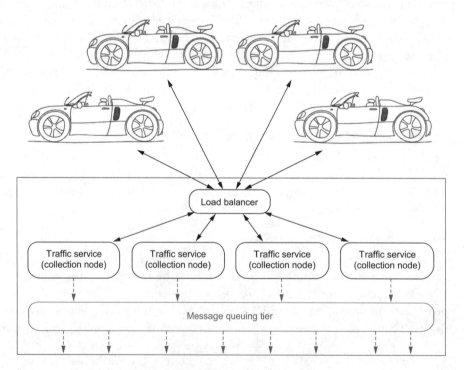

Figure 2.10   Vehicle and traffic service with a load balancer

We're using a load balancer here to be able to route requests from the vehicles to an instance of our service that's running. As instances are started or stopped based on demand, the running instances the load balancer routes requests to will change. We now have a pretty good idea of how we're going to scale our service and the protocol we're going to use with our clients.

2.2.2 Scaling the stream pattern

The meetup.com stream we used as an example in section 2.1.5 for our discussion of the stream interaction pattern has a fairly low velocity (less than 10 events per second). Obviously that's not the best example to help us think through how to scale a collection tier when using the stream interaction pattern. Instead, let's imagine that Google provided a public stream of all the searches being performed as they happen; according to internetlivestats.com (www.internetlivestats.com/one-second/#google-band) at this moment, that would result in approximately 46,000 search events per second. Remember, horizontal scaling is our goal when building each tier of our streaming system. With many streaming protocols, as you saw earlier when you consumed the meetup.com RSVP stream in your browser, there's a direct and persistent connection between the client (our collection tier) and the server (the service we request data from), as illustrated in figure 2.11.

In figure 2.11 you can see that three of the four nodes are idle because there's a direct connection between the search stream and the node handling the stream. To scale our collection tier, we have a couple of options: scale up the collection node

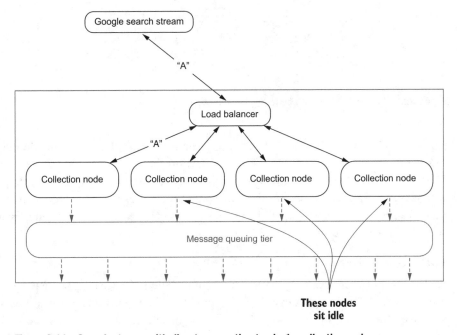

Figure 2.11 Search stream with direct connection to single collection node

that's consuming the stream, and introduce a buffering layer in the collection tier. These are not mutually exclusive, and depending on the volume and velocity of the stream, both may be required. Scaling up the node consuming the stream will get us only so far; at a certain point we'll reach the limits of the hardware our collection node is running on and won't be able to scale it up any further. Figure 2.12 shows what our collection tier looks like with the buffering layer in place.

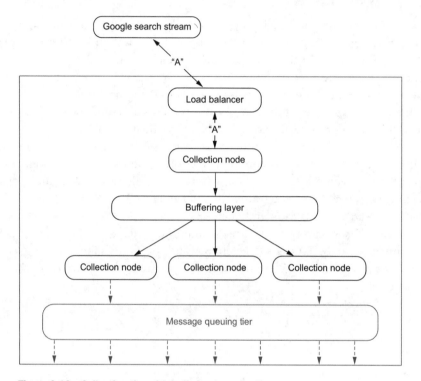

Figure 2.12 Collection tier with buffering layer in place

The key to being able to put a buffering layer in the middle lies in making sure no business logic is performed on the messages when they're consumed from the stream. Instead, they should be consumed from the stream and as quickly as possible pushed to the buffering layer. A separate set of collection nodes that may perform business logic on the messages will then consume the messages from the buffering layer. The second set of collection nodes can now also be scaled horizontally.

2.3   *Fault tolerance*

Regardless of the interaction pattern used, one thing is for sure: at some point one or more of our collection nodes will fail. The failure may be the result of a bug in our software, third-party software we rely on, or the hardware our service runs on. Regardless of the cause, our goal is to mask the failures and to improve the dependability of our collection tier. You may be wondering why we need to worry about this if we've done our job of horizontally scaling and increasing the redundancy of our tier. That's a fair question. The answer is quite simple: the message our collection tier receives from a client may not be reproducible. In essence, there may be no way for our collection tier to ask for the client to send us the data again, and in many cases no way for the client to do so even if our collection tier could ask.

Depending on your business, there may be times when it's okay to lose data, but in many cases it's not okay. This section explores the fault-tolerance techniques you can employ to ensure that we don't lose data and we improve the dependability of our collection tier. Our overarching goal is that when a collection node crashes (and it will), we don't lose data and can recover as if the crash had never occurred. To understand the areas we need to protect, look at figure 2.13, which shows the simplest possible collection scenario with the places we can lose data when the node crashes.

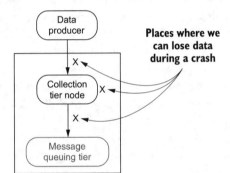

Figure 2.13 Collection scenario with our data-loss potential identified

The two primary approaches to implementing fault tolerance, *checkpointing* and *logging*, are designed to protect against data loss and enable speedy recovery of the crashed node. The characteristics of checkpointing and logging are not the same, as you'll soon see.

First, let's consider checkpointing. Numerous checkpoint-based protocols are available to choose from in the literature, but when you boil it down, the following two characteristics can be found in all of them:

- *Global snapshot*—The protocols require that a snapshot of the global state of the whole system be regularly saved to storage somewhere, not merely the state of the collection tier.
- *Potential for data loss*—The protocols only ensure that the system is recoverable up to the most recent recorded global state; any messages that were processed and generated afterward are lost.

What does it mean to have a global snapshot? It means we're able to capture the entire state of all data and computations from the collection tier through the data access tier and save it to a durable persistent store. That's what I'm talking about when

I refer to the *global state* of the system. This state is then used during recovery to put the system back into the last known state. The potential for data loss exists if we can't capture the global state every time data is changed in the system.

An example I'm sure you've seen before is autosave in popular document-editing software such as Microsoft Word or Google Docs. A snapshot is taken of the document as you are editing it, and if the application crashes, you can recover to the last check-point. If you're like many people, you've seen checkpointing and potential data loss in action when your word processing program crashed and your most recent edits were not saved.

When considering using a checkpoint protocol for implementing fault tolerance in a streaming system, keep two things in mind: the implications of the previously mentioned attributes and the fact that a streaming system is composed of many layers and many different technologies. This layering and the data movement make it very hard to consistently capture a global snapshot at a point in time, and that makes checkpointing a bad choice for a streaming system. But checkpointing is a valid choice if you're building the next version of HDFS or perhaps a new NoSQL data store. Given that checkpointing isn't a good match for a streaming system, we won't spend more time on these protocols. Even though they're not a good fit, they're fasci-nating to study. If you're interested in learning about them, I would encourage you to start with the great article by Elnozahy, En Mootaz, et al., "A Survey of Rollback-Recovery Protocols in Message-Passing Systems" (*ACM Computing Surveys* 34.3 (2002): 375–408).[2]

Turning our attention to the logging protocols, you have a variety to choose from. Reduced to their essence, you'll find that they all share the common goals of overcom-ing the expense and complexity of checkpointing and providing the ability to recover up to the last message received before a crash. Part of the complexity of checkpointing that's eliminated is the global snapshot, and therefore the management and generation of the global state. In the end, the goals of the logging technique manifest themselves in the basic idea that underpins all of the logging techniques: *if a message can be replayed, then the system can reach a global consistent state without the need for a global snapshot.*

This means that each tier in the system independently records all messages it receives and plays them back after a crash. Implementing a logging protocol frees us from worrying about maintaining global state, enabling us to focus on how to add fault tolerance to the collection tier. To do this we're going to discuss two classic tech-niques, *receiver-based message logging* (RBML) and *sender-based message logging* (SBML), and an emerging technique called *hybrid message logging* (HML). Along the way we'll also discuss how and why we can use these with our collection tier.

Before moving on to discuss these techniques, look at figure 2.14, which illustrates how they fit together and what data we're trying to protect.

Figure 2.14 shows a single collection tier node that's receiving a message, performing some logic on it, and then sending it to the next tier. As their names imply, receiver-based

---

[2] The article can be downloaded from http://mng.bz/vUz2.

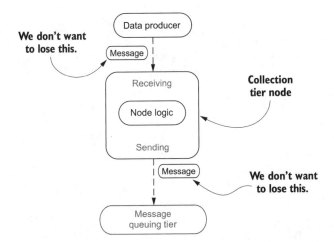

Figure 2.14 High-level overview of receiver-based and sender-based message logging

logging is concerned with protecting the data the node is receiving, and sender-based logging is concerned with protecting the data that's going to be sent to the next tier. Imagine your business logic being sandwiched between two layers of logging, one designed to capture the data before it's changed and one to capture it before it's sent to the next tier. If you're thinking that this is a lot of potential overhead and overlap, in some cases it may be, and this is where HML aims to strike a balance between RBML and SBML. With that frame of reference, let's start our discussion with RBML.

2.3.1 Receiver-based message logging

The RBML technique involves synchronously writing every received message to stable storage before any action is taken on it. By doing that, we can ensure that when our software crashes while handling the message, we already have it saved and upon recovering we can replay the message. Figure 2.15 illustrates how our collection node changes with the introduction of RBML.

In figure 2.15 the message flows from step 1 to step 5; this shows the happy path when there is no failure. We'll walk through the recovery side of it shortly, but first let's review the flow:

1 A message is sent from a data producer (any client).
2 A new piece of software we wrote for the collection node, called the RBML logger, gets the message from the data producer and sends it to storage.
3 The message is written to stable storage.
4 The message then proceeds through to any other logic we have in the node; perhaps we want to enrich the data we're collecting, filter it, and/or route it based on business rules. The important thing is we are recording the data as soon as it is received and before we do anything to it.
5 The message is then sent to the message queuing tier, the next tier in the streaming system.

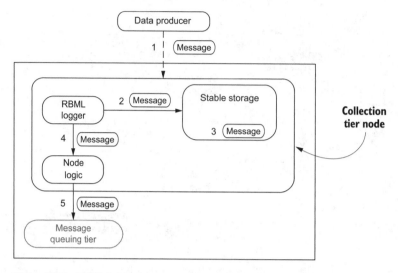

Figure 2.15   RBML implemented for the simple collection node showing the happy path

It's important to point out that depending on the type of stable storage used, steps 2 and 3 may negatively impact the throughput performance of our collection node, sometimes noted as one of the drawbacks to logging protocols. The hybrid message logging technique discussed in section 2.3.3 helps address some of those concerns. For now, we'll keep it simple—at the end of the day the simplicity and recoverability of using RBML for our collection node wins.

Now that you understand how the data flows during normal operation, look at figure 2.16, which shows what the recovery data flow looks like.

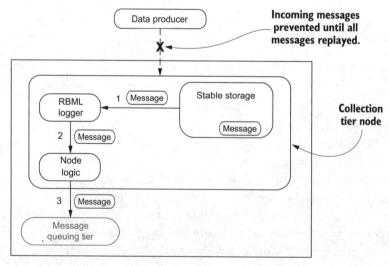

Figure 2.16   The recovery data flow for RBML

In figure 2.16 there are a couple of things to call out. First, once the crash occurs, all incoming messages to this collection node are stopped. Because you'll have more than one collection node and they'll be behind a load balancer, you would take this node out of rotation. Next, the RBML logger reads the messages that have not been processed from stable storage and sends them through the rest of the node logic as if nothing has happened. Lastly, after all pending messages are processed, the node is considered restored and can be put back into rotation, and the data flow resumes, as in figure 2.15.

2.3.2 Sender-based message logging

The SBML technique involves writing the message to stable storage before it is sent. If the RBML technique logs all messages that come in the front door of our collection node to protect us from ourselves, then SBML is the act of logging all outgoing messages from our collection node before we send them, protecting ourselves from the next tier crashing or a network interruption. Figure 2.17 shows the data flow for SBML.

Now that you understand RBML and in particular the data flow, I suspect that the data flow for SBML as depicted in figure 2.17 seems fairly reasonable to you through step 5. One difference is that with RBML we are recording the message as soon as it is received before we do anything to it, and with SBML we are recording it before we send any data to the next tier. The data recorded by an RBML logger is the raw incoming data, and the data recorded by the SBML logger is after our node logic executes (remember, we may have augmented the data in some way) and before we send it on.

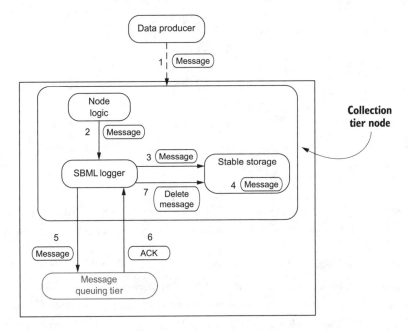

Figure 2.17   The normal execution data flow for SBML

Besides this nuance there's a little wrinkle we'll need to deal with during recovery. During recovery, how do we know whether the next tier has already processed the message we're replaying? There are several ways to handle this. One, shown in figure 2.17, is that we use a message queuing tier that returns an acknowledgment that it received the message. With that acknowledgment in hand, we can either mark the message as replayed in stable storage or delete it from stable storage because we no longer need to replay it during recovery. If the technology you choose for your message queuing tier doesn't support returning an acknowledgment of any sort, then you may be forced into a situation where if no error occurs when sending the message to the message queuing tier, steps 6 and 7 will result in you deleting the message from stable storage.

The recovery data flow as illustrated in figure 2.18 is a little more complex than how we handled recovery with the RBML.

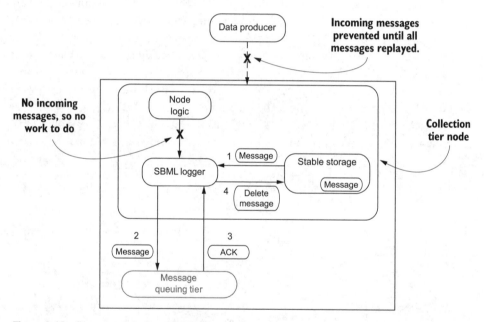

Figure 2.18   Recovery data flow for SBML

I think you'll agree that the recovery data flow for SBML is only marginally more complex than the RBML workflow, but it shouldn't look too foreign to you.

2.3.3   *Hybrid message logging*

See code listings 9.1 and 9.8

If we stopped right now, we'd have two solutions that we can put in place to address our data loss concerns and dependability: RBML to handle incoming messages and SBML to handle outgoing messages. As mentioned, writing to stable storage can negatively impact our collection node's performance. Implementing both RBML and SBML means we're writing to stable storage at least twice during normal execution. Some may argue that we'll be doing more logging than processing of data; figure 2.19 suggests they may not be far off.

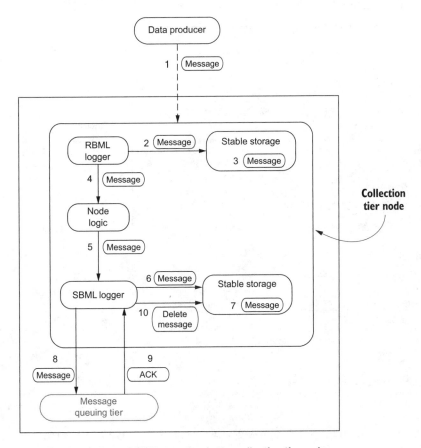

Figure 2.19   RBML and SBML together in the collection tier node

To help with this, hybrid message logging was designed to reduce the impact of message logging during normal processing. To accomplish this, the best parts of RBML and SBML are used at the cost of minimal additional complexity. HML is also designed to provide the same data-loss protection and recoverability found in RBML, SBML, and other logging techniques. There are several ways to implement HML; one common approach is illustrated in figure 2.20.

It's apparent when comparing the data flow in figure 2.19 with both RBML and SBML to the HML data flow in figure 2.20 that the HML approach is slightly less complex. Several factors contribute to this simplification. The first one, which may not come as a surprise, is that the two stable storage instances have been consolidated. This is a minor change, but it allows you to reduce the number of moving parts. The

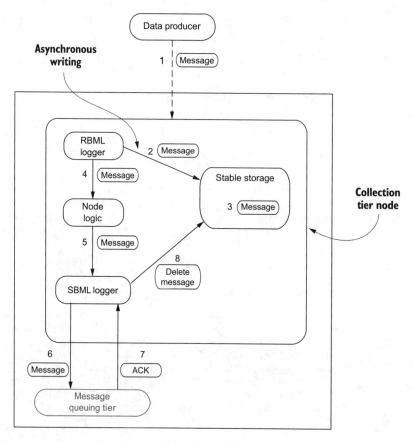

Figure 2.20   HML sample data flow

second change, writing to stable storage asynchronously, has a subtle difference. Arguably this has a more profound impact on the implementation complexity and performance. The complexity comes from making sure you are correctly handling any errors that happen, and the performance comes from using the multi-core world we live in to perform more than one task at a time.

The rest of the data flow should be routine for you by now. If you feel comfortable with the additional complexity and have a choice of implementing HML or standard RBML and SBML, you should implement HML because it will reduce the performance impact of message logging while still providing all the fault tolerance and safety of RBML and SBML. If you're interested in learning more about HML, a great place to start is an article on hybrid message logging by Hugo Meyer and others.[3] In

---

their research, they showed how they were able to achieve a 43% overhead reduction over the RBML approach.

2.4 A dose of reality

Here's a funny little story to put some of this scaling and fault tolerance into perspective. One time I was working on a streaming system that was populating fancy dashboards for marketers. It had all the bells and whistles—scaling, fault tolerance, monitoring, alerting—the whole nine yards. We had to have all of this and could not lose any data, because our customers wouldn't accept a solution that didn't have complete data. Once this system was running in production, I was curious as to how well our web-based dashboards that consumed our stream via WebSockets were keeping up. Well, come to find out, many of our customers were only able to keep up with about 60% of the stream that was being sent to them; the other 40% of the data was being dropped because they couldn't read it fast enough. When I mentioned this to coworkers, they were shocked and somewhat in disbelief because our customers and business folks loved what they were seeing.

It put things in perspective: the dashboards we produced were showing a picture of our customers' business that was not distorted by the missing data. To me, this was like the difference between high-end HDTV and mid-level HDTV—sure, the quality of the picture may be slightly better, but the picture doesn't change. I'm not implying that you don't need to worry about scaling or fault tolerance, but it's good to keep things in perspective and then reflect on the difference between "we must have *xyz* features" and reality.

2.5 Summary

We've covered a lot of ground in this chapter, exploring the various aspects of collecting data for a streaming system from interaction patterns through scaling and fault-tolerance techniques.

Along the way you

- Learned about the collection tier
- Developed an understanding of various collection patterns
- Had a chance to interact with a live stream
- Learned how to think about scaling your collection tier
- Learned about common fault-tolerance techniques

Transporting the data from collection tier: decoupling the data pipeline

This chapter covers

- Understanding the need for the message queuing tier
- Understanding message durability
- Accommodating offline consumers
- Understanding message delivery semantics
- Choosing the right technology

So far I've talked about the role of handling the incoming data, not the output of data, from the collection tier. This chapter focuses on transporting data from the collection tier to the rest of the streaming pipeline. Although I may mention the collection and analysis tiers in the discussion, the discussion will only be concerned with getting messages from or to those tiers via the message queuing tier. Figure 3.1 shows our streaming architecture with this focus in mind.

After completing this chapter you will have a solid understanding of why we need a message queuing tier, what the core features of the common products used in this tier are, and how to determine which features are important for your streaming data system.

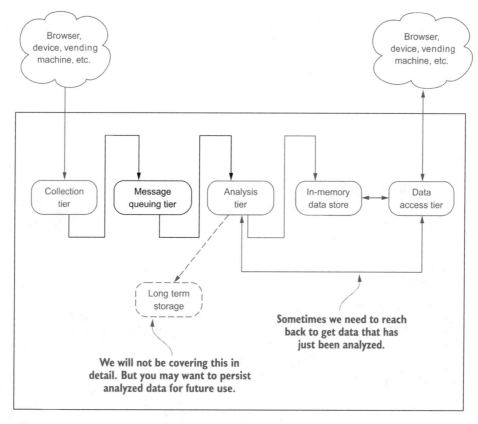

Figure 3.1　The message queuing tier with its input and output as the focus

3.1　*Why we need a message queuing tier*

If you're new to building data pipelines, and in particular streaming systems, you may look at figure 3.1 and think, "Okay, I figured the output from the collection tier went to the message queuing tier and then the data magically flows to the analysis tier, so what's the big deal? And why do we need this message queuing tier at all?" Those are great questions, so let's imagine for a second that our streaming architecture did not have the message queuing tier as part of it. Figure 3.2 shows a redrawn architecture without this tier.

We may be tempted to say this looks simpler and things should work fine, but don't give in to this desire for what appears to be a simplification of the architecture. Sure, you may be able to bring up an entire streaming system on a single machine and have each layer directly call the next, but your streaming system will span across many machines. When designing a software system, one desirable quality to strive for is the decoupling of the various components. With a streaming system we want the same: to decouple the components in each tier—but more importantly, to decouple the tiers

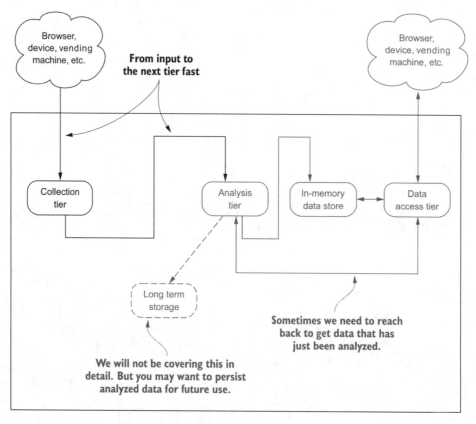

Figure 3.2   From the collection tier straight to the analysis tier

from each other. At the heart of a streaming system, like any distributed system, is the communication between the numerous machines that compose the system. If you look up *interprocess communication* in the literature[1] you will find numerous models; this chapter focuses on the *message queuing model*. By adopting this model, our collection tier will be decoupled from our analytics tier. This decoupling allows our tiers to work at a higher level of abstraction, by passing messages and not having explicit calls to the next layer. These are two good properties to have in any system, let alone a distributed streaming one. As you will see in this and upcoming chapters, this decoupling of the tiers provides wonderful benefits.

3.2   Core concepts

Let's look at the features of a message queuing product that are critical to the success of our streaming system. But first let's get one more formality out of the way—our use of the term *message queuing*. I use such terms broadly to encompass the spectrum

---

[1]   An overview of inter-process communication, https://en.wikipedia.org/wiki/Inter-process_communication.

of messaging services, from the traditional message queuing products (RabbitMQ, ActiveMQ, HornetQ, and so on) to the newer takes on messaging found in NSQ, ZeroMQ, and Apache Kafka. Apache Kafka has grown to embody more than a message system, and I'm going to focus on the fact that it lets us publish and subscribe to streams of records. When you consider this functionality, it is similar to a message queue or enterprise messaging system.

This section discusses where this tier fits in the larger streaming architecture and talks about the core features to consider when selecting a message queuing product—only the ones you want to pay attention to when designing a streaming system. Armed with this information, you will be able to objectively think about the problem we are trying to solve in the context of what is important to your business, making the selection of the correct tool easier.

Before diving into the core features, let's make sure we have an understanding of the components of a message queuing product and how they map to our streaming architecture.

3.2.1 *The producer, the broker, and the consumer*

In the message queuing world there are three main components: the *producer*, the *broker*, and the *consumer*. Each plays an important role in the overall functioning and design of the message queuing product. Figure 3.3 shows how they fit together in their simplest form.

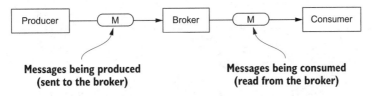

Figure 3.3   **The three core parts to a message queuing system**

You can see that the producer and the consumer have jobs that closely match their names: the producer produces messages, and the consumer consumes messages. Note in figure 3.3 that the term *broker* is used and not *message queue*. Why the change? Well, it's not so much a change as it is an abstraction, and this abstraction is important because a broker may manage multiple queues. Figure 3.4 shows that the message queue is alive and well but is abstracted away by the broker.

Now the data flow should start to make more sense. If you follow the flow from left to right in figure 3.4, you will see the following steps taking place:

- The producer sends a message to a broker
- The broker puts the message into a queue
- The consumer reads the message from the broker

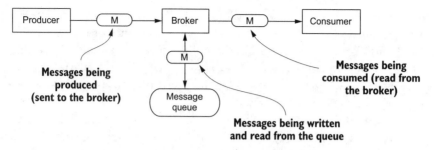

Figure 3.4   The broker with the message queue being shown

To put this in perspective, figure 3.5 shows what overlaying these terms and pieces onto our streaming architecture looks like.

You probably would agree that this seems pretty simple and straightforward, but as the saying goes, the devil is in the details. It is to these details—the subtle interactions between the producer, broker, and consumer, as well as various behaviors of the broker—that we will now turn our attention.

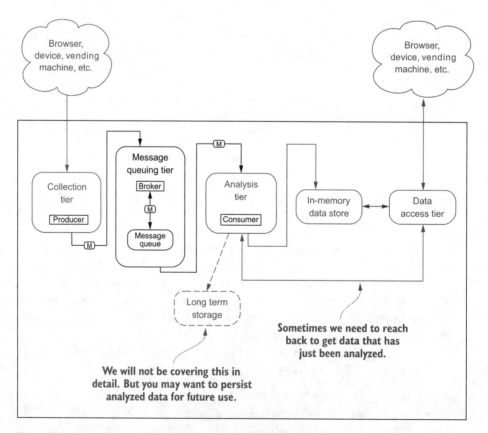

Figure 3.5   Streaming architecture with message queuing tier components shown in context

3.2.2   *Isolating producers from consumers*

As mentioned, one of our goals is to decouple the different tiers in the system. The message queuing tier allows us to do this with the collection and analysis tiers. Let's explore several reasons why this is a quality we strive for and the benefits we get from it.

Depending on your business need and the design of your streaming system, you may find yourself in a situation where the producer (collection tier) is generating messages faster then the consumer (analysis tier) can consume them. Often this is because your analysis tier is more processing-intensive than the collection tier and may not be able to process the data as quickly, which presents the following challenge: What if our consumers can't consume data fast enough from the collection tier? How do we prevent our analysis tier from being overwhelmed by surges in the events produced in the collection tier?

To me this conjures an old cartoon picture of a hose with the end plugged and the water spigot opened—it starts to swell and eventually explodes from the backpressure. Taking that example to our realm of data, figure 3.6 shows a time-lapse of the data flowing from the collection tier to the analysis tier.

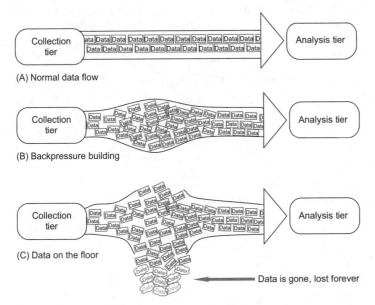

Figure 3.6   The three stages of data flowing without a message queue. We don't want step C.

Let's take figure 3.6 step-by-step:

- *Step A*—This looks pretty normal and is what we would like to see.
- *Step B*—We can tell something is not quite right—backpressure is building.
- *Step C*—Our data pipe broke under pressure, and data is now virtually dropping onto the floor and is gone forever.

Ouch! This is not a good situation because we are now losing data, and for some businesses that can be catastrophic. At first blush, you may think this is a consumer problem, and all we have to do is add more consumers or make them faster so they can keep up and life will be good. But this is not a consumer problem at all; it is perfectly acceptable in many use cases for consumers to read slowly or be offline from time to time. For example, consumers may only want to read all the messages on an hourly basis to support a batch-processing use case; they will be offline from time to time and then connect and read the last hour's worth of data.

Remember, not all message queuing systems provide this type of producer flow control, leaving it up to you, the application developer, to control the rate at which your collection tier is producing messages. If you don't, you may overwhelm your consumer and the broker. The ability to support a consumer reading messages slowly or being offline from time to time is provided by message queuing products that support durable messaging.

3.2.3   *Durable messaging*

Why do you need to worry about durable messages and offline consumers in a book about building streaming data systems? Great question! Let's look at what else we get in exchange for using a message queuing product that supports offline consumers. Imagine that the data centers for your business are geographically dispersed. You have two data centers: one in Amsterdam, the Netherlands, and the other in San Diego, CA, as shown in figure 3.7.

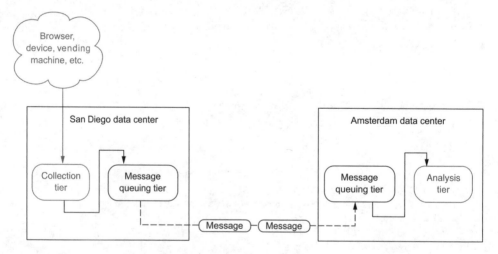

Figure 3.7   Two data centers with data flowing between them

In the San Diego data center you have the collection tier running, and in the Amsterdam data center you're running the analysis tier. I haven't talked about the analysis tier yet, but it needs the data from the collection tier. All right, you are collecting data in San Diego and analyzing it in Amsterdam. Things are running smoothly and business is good. But as luck would have it, right as you were about to leave for the weekend on a beautiful Friday afternoon, a construction worker accidently put a backhoe through a fiber optic line, cutting off communication between your data centers, as illustrated in figure 3.8.

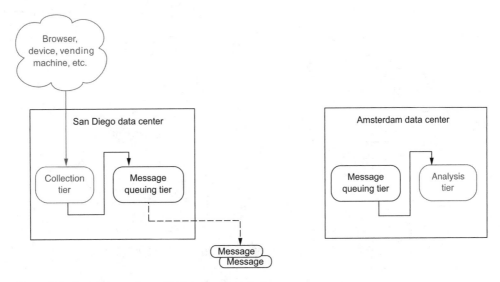

Figure 3.8   Two data centers with data flowing into the ocean

The telecom company that owns the fiber line says its best guess is that it may take two or three days to repair it. What would be the impact on your business? How much data can your business tolerate losing from your collection tier? If you couldn't tolerate losing days of data, make sure you choose a message queuing technology that can persist messages for the long term. Figure 3.9 shows how durable messaging fits in with this tier and some of the types you may find.

Having durable messages not only provides a degree of fault tolerance—and therefore disaster recovery—it allows for the offline consumer scenario mentioned earlier.

Imagine you've built a real-time traffic-routing system that allows people driving around any city to use your smartphone app to get updates and be re-routed based on up-to-the-moment traffic conditions. Three months pass, and now your business wants to offer a historical traffic-replay product that lets a user pick a city and replay the traffic data for a given day, week, or month. If your architecture is similar to figure 3.10, then as the analysis tier consumes messages they are discarded from the message queue—in essence, they are gone, and you can't provide your historical traffic replay.

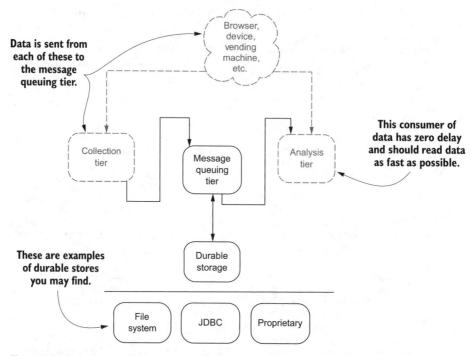

Figure 3.9   Durable messages—where they fit and how they may be stored

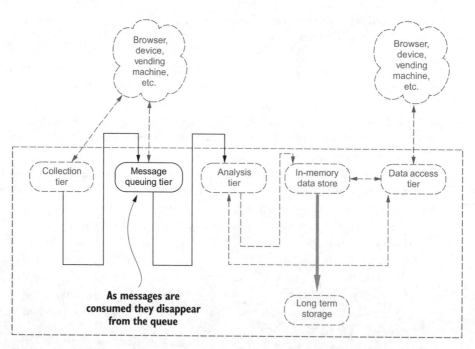

Figure 3.10   Transient messages get discarded after the analysis tier consumes them.

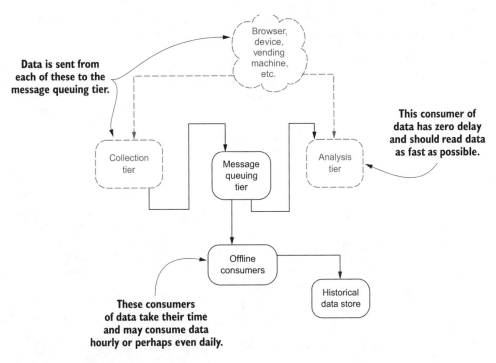

Figure 3.11 Offline consumers persisting data for historical reporting/analysis

To solve this problem, you need an architecture more like figure 3.11.

If your business case requires supporting offline consumers, or you want to be sure your producers and consumers can be completely decoupled, look for a product that supports durable messaging.

3.2.4 *Message delivery semantics*

Remember, a producer sends messages to a broker, and the consumer reads messages from a broker. That's a pretty high-level description of how message delivery works. Let's go deeper and explore the common semantic guarantees found in messaging products. The following are the three common semantic guarantees you will run into when looking at message queuing products:

- *At most once*—A message may get lost, but it will never be reread by a consumer.
- *At least once*—A message will never be lost, but it may be reread by a consumer.
- *Exactly-once*—A message is never lost and is read by a consumer once and only once.

If you had to pick the one you wanted, which would it be? If you said *exactly-once* you're not alone. In fact, most people want a system in which messages are never lost and each message is delivered to a consumer once and only once. Who wouldn't want that? If only it were that simple—such a system comes with caveats and risks. Figure 3.12 shows possible points of failure.

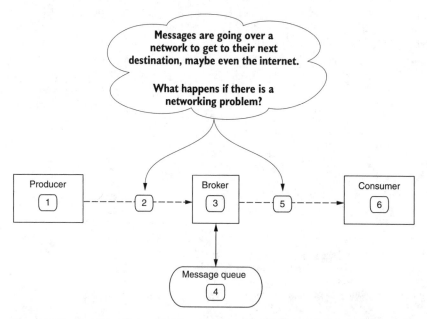

Figure 3.12   Possible points of failure that need to be considered

Wow! It seems as if almost every spot in the diagram is a possible point of failure. But don't worry, it's not all doom and gloom. Let's walk through them and understand what the risks are and what each numbered item in figure 3.12 means:

1 *Producer*—If the producer fails after generating a message but before sending it over the network to the broker, we will lose a message. There is also a chance that the producer may fail waiting to hear back from the broker that it did receive the message, and the producer after it recovers may send the same message a second time.

2 *The network between the producer and broker*—If the network between the producer and the broker fails, the producer may send the message, but the broker never receives it or the broker does receive it but the producer never gets the response acknowledging it. In both cases the producer may send the same message a second time.

3 *Broker*—If the broker fails with messages that are still held in memory and not committed to a persistent store, we may lose messages. If the broker fails before sending an acknowledgment to the producer, the producer may send the message a second time. Likewise if the broker tracks the messages consumers have read and fails before committing that information, a consumer may read the same message more than once.

4 *Message queue*—If the message queue is an abstraction over a persistent store, then if it fails trying to write data to disk we may end up losing messages.

5 *The network between the consumer and broker*—If the network between the consumer and the broker fails, the broker may send a message and record that it was sent, but the consumer may never get it. From the consumer side, if the broker waits for the consumer to acknowledge receipt of a message but that acknowledgment never gets to the broker, it may send the consumer the same message a second time.

6 *Consumer*—If the consumer fails before being able to record that it processed a message, either by sending an acknowledgment to the broker or to a persistent store, it may request the same message from the broker. Another twist here is the case where there are multiple consumers and more than one of them reads the same message.

I know that is a lot to consider and it may seem a little overwhelming, but don't worry. This won't be the last time we discuss these types of semantics. In the context of a message queuing system, we need to keep these failure scenarios in our back pocket so that when a messaging system claims to provide exactly-once deliver semantics, we can understand whether it truly does. As with so many things, the choice of which technology to use in this case involves various tradeoffs, as listed in table 3.1.

Table 3.1 Tradeoffs of a message queuing system

Less complexity, faster performance, and weaker guarantees	vs.	More complexity, a performance hit, and a strong guarantee

Where to compromise should be based on the business problem you're trying to solve with the streaming system. If you're building a streaming web analytics product, missing a message here or there is not going to have much of an impact on your product. Conversely, for a streaming fraud detection system, missing a message can have an undesirable effect.

As you look at messaging systems, you may find that the one you want to use doesn't provide exactly-once guarantees, such as Apache Kafka and Apache ActiveMQ. But don't despair. Often the system will provide enough metadata about the messages that you can implement exactly-once semantics with some coordination between producer(s) and consumer(s). Let's explore the techniques we would use to solve this problem. Figure 3.13 illustrates them graphically.

Figure 3.13 identifies the producer and consumer techniques. Let's talk about those in more detail:

- *Do not retry to send messages*—This is the first technique we must use. To do this, you need to have in place a way to track the messages your producer(s) sends to the broker(s). If and when there is no response or a network connection is interrupted between your producer(s) and the broker(s), you can read data from

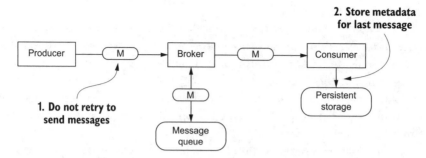

Figure 3.13   The two ways to have exactly-once semantics if the messaging system doesn't provide it

the broker to verify that the message you didn't receive an acknowledgment for was received. By having this type of message tracking in place, you can be sure your producer only sends messages exactly-once.

- *Store metadata for last message*—The second technique we must use involves storing some data about the last message we read. The metadata you store will vary by messaging system. If you're using a JMS-based system, you may store the JMS-MessageID; if you're using Apache Kafka, you would store the message offset. In the end, what you need is data about the message so you can be sure your consumer doesn't reprocess a message a second time. Figure 3.13 shows the metadata being stored in a persistent store. One thing to take into consideration is what to do if there's a failure storing the metadata.

If you implement these two techniques, you can guarantee exactly-once messaging. You may not have noticed during this discussion, but by doing so you also get two nice little bonuses—sorry, not that type of bonus. I was thinking more about the data quality and robustness of your system. Look again at figure 3.13 and the subsequent discussion. What do you think the bonuses are? There may be more, but the ones I was thinking of are message auditing and duplicate detection.

From a message auditing standpoint, you're already going to keep track of the messages your producer sends via metadata. On the consumer side you can use this metadata to keep track of not only messages arriving but also the max, min, and average time it takes to process a message. Perhaps you can identify a slow producer or slow consumer. Regarding duplicate detection, we already decided that our producer was going to do the right thing to make sure a message was only sent to a broker one time. On the consumer side, we said it was going to check to see whether a message has already been processed.

One extra thing to keep in mind: on the consumer side, don't merely keep track of metadata related to the messaging system (some will expose a message ID of some sort so you know if you processed a message by the same ID). Also be sure to keep track of metadata that you can use to distinctly identify the payload of a message. This will make the de-duplication of messages and auditing of data easier.

Now you know how to ensure exactly-once semantics, and you're on your way to providing message auditing and detecting message duplication. In this book you'll run into these concepts again, and you may see other ways to apply message auditing through the entire streaming architecture.

3.3 Security

Up until now we have only been concerned with making sure we don't lose data or overwhelm the broker or consumers and that we understand how we can recover from a failure. This is all good, because now we know how to provide a robust message queuing tier. We have one more hurdle to cover: security. In this day and age, not only is it important that we secure the data in-flight and at rest, but that we also can ensure that a producer is allowed to produce messages and a consumer is allowed to consume them. Figure 3.14 shows all the places we need to think about when securing the message-queuing tier.

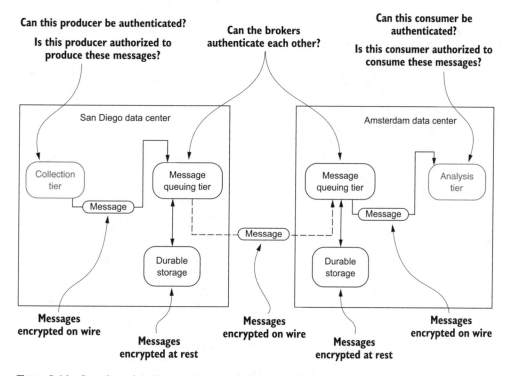

Figure 3.14 Security points that need to be considered for the message queuing tier

It may seem as if we have a daunting task ahead of us to secure the message queuing tier. At a minimum, you need to think through these aspects, and if you work with a security group, it would be good to engage them as you undertake securing this and the other tiers. At a high level I'm sure they all make sense, but as I've mentioned, the devil is in the details, including figuring out how to work around some of the limitations when the message-queuing product you chose doesn't support everything you may need. There are many great resources to consult regarding security as you proceed further. A good place to start for in-depth coverage of security in distributed systems is Ross Anderson's *Security Engineering: A Guide to Building Dependable Distributed Systems* (Wiley, 2008).

3.4 Fault tolerance

Now that our collection and analysis tiers are isolated and we're sending messages through the message queuing tier, you need to understand what happens to the data when things go wrong, and it's not a matter of *if* things will go wrong, but *when*. For the multiple data center architecture we've been talking about, how many places can you identify where we may lose data? Look at figure 3.15 and compare notes.

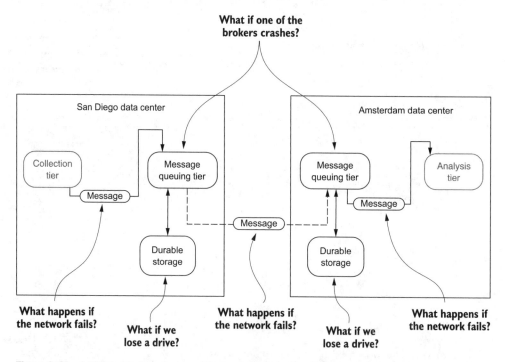

Figure 3.15 A high-level view of where things can go wrong

In figure 3.15 several callouts should be familiar to you from chapter 2, notably the producer having to handle network failures. The network failing or being unavailable between data centers is a reality that can often be mitigated with a broker that uses durable storage. If the message queuing product that meets your business needs doesn't use a durable store, you will either need to live with data loss or find another way to mitigate the risk of losing network connectivity. The callout in the figure on the consumer side regarding the network failing while the analysis tier is consuming the message is covered in chapter 4.

Let's dig a little deeper into what can go wrong in this tier. Figure 3.16 shows a zoomed-in view of the message queuing tier.

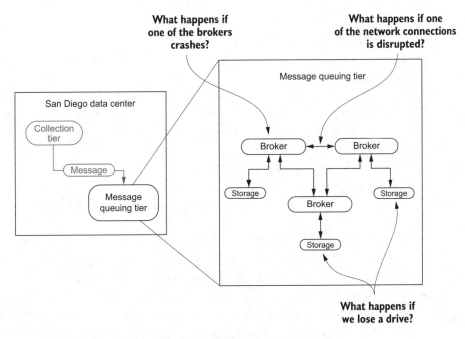

Figure 3.16   Zoomed in view of brokers and what can go wrong

In this zoomed-in view you can see we have three brokers—the number of brokers isn't as important as what's going on between them and where things can fail. When it comes to particular message queuing products, the number of brokers you deploy will matter, and you should have a thorough understanding of how best to utilize the product you chose. For now the goal is to make sure you know where things can break and what questions you may need to answer when evaluating various products.

Let's now tackle the questions identified in figure 3.16. These questions and the following discussion are not exhaustive but should serve as a good starting point when thinking through this problem in greater detail.

- *What happens if a broker crashes?*—This is an interesting problem. If the broker uses durable storage, we would hope that the only data at risk are those messages that are in memory when the broker crashes. There are ways we can mitigate this risk:
  - You can apply the lessons of chapter 2: to guarantee the message was delivered, our producer should wait for an acknowledgment that the data was indeed written to disk.

– You can have the message queuing product replicate the message to more than one broker. In this case you still have risk, but it's been reduced significantly because now two or more brokers must crash at the same point in time before the messages are written to disk.

– You can configure the broker to hold as little data in memory as possible. This method may have performance implications, so you are trading potential performance for durability.

- *What happens if the network between two brokers is interrupted?*—For many of the message queuing systems that provide replication, your data is safe because it is on more than one broker. Also, once the network connection is restored, the broker will rejoin the cluster and synchronize the messages it missed from other brokers. Think through the following related questions with the message-queuing product you chose:

  – Is a different broker chosen as the new replica?

  – What happens if the network connection is restored?

  – Is there an ability to configure the delay before the network is deemed to be "down"?

  – What happens to the data if the network connection isn't restored?

  – What happens to the data if a producer was in the middle of sending a message to the broker that got severed from the cluster?

- *What happens if we lose a drive?*—Although this has the potential to be a drastic loss, many operations people are used to dealing with it. In the case of a message-queuing product, if multiple brokers are involved, then these questions follow:

  – Are there replicas of the data that was lost?

  – What if there is data that was being replicated and it wasn't written to disk when the drive was lost—is that data lost?

  – How can you recover a broker?

Armed with these questions, I hope you will be able to apply them when assessing the fit of a particular product for your business problem.

3.5 Applying the core concepts to business problems

Now that we have covered the core concepts to keep in mind regarding the message queuing tier, let's see if we can apply them to different business scenarios.

FINANCE: FRAUD DETECTION

Paul's company provides real-time fraud-detection services that help detect fraud as it is happening. To do this he needs to collect credit card transactions from all over the web as they are occurring, apply some pretty cool algorithms against the data, and then send back approved or declined messages to customers while a purchase is happening. Bearing in mind the core concepts we've discussed, table 3.2 lists things that may come up when designing the architecture for Paul's business.

Table 3.2   Fraud detection scenario questions

Question	Discussion
What would be the impact to Paul's business if the communication between the collection tier and analysis tier were interrupted for an extended period of time?	The impact would be catastrophic. Paul's business would not be able to offer its service and may cause a detrimental impact to his customers' businesses.
How many days' worth of data can Paul's business tolerate losing?	Zero. In fact, I would argue that given the nature of Paul's business and the type of data he's dealing with, losing data is not an option.
Would you anticipate that Paul would need to store historical data?	I would expect that at least one customer or an executive in Paul's business has asked to see a report detailing how their service has performed over time.
What type of message delivery semantics does Paul's streaming system need?	I would expect that his business needs exactly-once semantics. Without that, he may miss a message, and therefore miss a fraudulent transaction. Could he get by with at least once? Perhaps. It may make the consumer more complex, but it is possible.

That was fun. Now let's take two more totally different businesses and see if you can answer the questions for them.

INTERNET OF THINGS: A TWEETING COKE MACHINE

Frank's business owns thousands of Coke vending machines and would like to make them social. He wants his machines to tweet and send push notifications with special offers to consumers who are geographically close. As if that weren't enough, there's a twist: If the closest vending machine doesn't have stock to offer, it should recommend the next closest vending machine that can offer the customer a deal. How would you answer the questions posed in table 3.3? I've added some things to think about in the discussion column. Considering the totally different business requirements, don't be surprised if your answers are quite different from the previous example. The idea is to spend time thinking through the problem.

Table 3.3   Internet of Things scenario questions

Question	Discussion
What would be the impact to Frank's business if the communication between the collection tier and analysis tier were interrupted for an extended period of time?	Obviously, Frank would lose money or could lose money if a social vending machine has a significant impact on his revenue. What if he also used this for inventory management? Would the impact be larger?
How many days' worth of data can Frank's business tolerate losing?	That may depend on how he handles inventory and the popularity of the machines. Perhaps this is locale-specific.
Would you anticipate that Frank would need to store historical data?	Perhaps—if he wanted to do reporting on how well his social vending machines are performing.

Table 3.3   Internet of Things scenario questions *(continued)*

Question	Discussion
What type of message delivery semantics does Frank's streaming system need?	In this case, at least once should suffice, although I think you could make an argument for at most once. What do you think would be the impact if a message were processed more than one time? What if a message was missed entirely?

E-COMMERCE: PRODUCT RECOMMENDATIONS

Rex runs a high-end fashion e-commerce business and is trying to increase the conversion rate on his site. He feels that perhaps social influence is one way to do this. To support this effort, Rex has asked you to architect a system that will let him show visitors to his site what other people have recently added to their cart or purchased. If I'm looking at a pair of jeans, and other customers on the site added shoes along with the same jeans to their cart or purchased a shirt with the same jeans, I should see the shoes and shirt recommended for me to buy along with the jeans. Imagine the messaging being something like this: "Ten other people just purchased these [product name] with the jeans you're looking at." Or "Five people also added this shirt to their cart along with these jeans." To summarize, the requirements we are trying to fulfill are:

- Keep track of all purchases in real time
- Keep track of all shoppers carts in real time
- Show on every product page the products recently purchased together
- Show on every product page the related products current shoppers have in their carts right now

With our mission in mind, answer the questions in table 3.4 so we can design the correct system for Rex. I have added input to the discussion column that may aid in your analysis.

Table 3.4   E-commerce requirements questions

Question	Discussion
What would the impact be to Rex's business if the communication between the collection tier and analysis tier were interrupted for an extended period of time?	Not being able to have the marketing opportunities to up-sell and cross-sell would cause this feature to not function. Imagine if you went to Amazon and were not shown the product recommendations. It would be hard for us to quantify the monetary value for Frank, but for sure it would be lost opportunity and lost sales.
How many days' worth of data can Rex's business tolerate losing?	Perhaps we could provide an old-fashioned batch-based recommendation system as a backup, but the freshness of the data may be an issue.
Would you anticipate that Rex would need to store historical data?	Perhaps for reporting purposes and for refinement of any algorithms that we develop.

Table 3.4 E-commerce requirements questions *(continued)*

Question	Discussion
What type of message delivery semantics does Rex's streaming system need?	Having at least once should suffice for this use case. Exactly-once is overkill. Do you think we would need at most once?

3.6 *Summary*

See section 9.2.1

In this chapter we've explored how to decouple the data being collected from the data being analyzed by using a message queuing tier in the middle. During this exploration we did the following:

- Learned why we need the message queuing tier
- Developed an understanding of message durability
- Learned how to accommodate offline consumers
- Learned about the different message delivery semantics
- Learned how you can choose the right technology for your business problem

At this point we've developed a good understanding of how to decouple the data being produced by our collection tier from the analysis tier. As you move through the other chapters, you'll see some of the terms and concepts popping up again, so don't worry if this is a bit overwhelming right now—it won't be by the last time we talk about message delivery semantics. If you're interested in learning more about the numerous aspects of messaging systems and the variety of integration patterns, see Gregor Hohpe and Bobby Woolf's *Enterprise Integration Patterns* (Addison-Wesley, 2003).

Now let's get ready to have fun with the data we've collected. The next chapter takes us through the analysis tier—our message consumer. Ready? Let's go.

Analyzing streaming data 4

This chapter covers

- In-flight data analysis
- The common stream-processing architecture
- Key features common to stream-processing frameworks

In chapter 3 we spent time understanding and thinking through the importance of the message queuing tier. That tier is designed to gather data from the collection tier and make it available to be moved through the rest of the streaming architecture. At this point the data is ready and waiting for us to consume and do magic with. In this chapter you're going to learn about the analysis tier. Our goal is to get to know the underlying principles of this tier, and in chapter 5 we'll dive into all the ways to use this tier to perform magic on the data. With that frame of reference in mind, consult our navigational aid in figure 4.1 to make sure you're oriented with respect to the flow of data.

Notice in figure 4.1, unlike in chapter 3 which discussed the input and output of the data, here we're only going to concern ourselves with the input. We'll hold off on talking about where the data goes from this tier until the next chapter. After finishing this chapter you'll know the core concepts found in all the modern tools

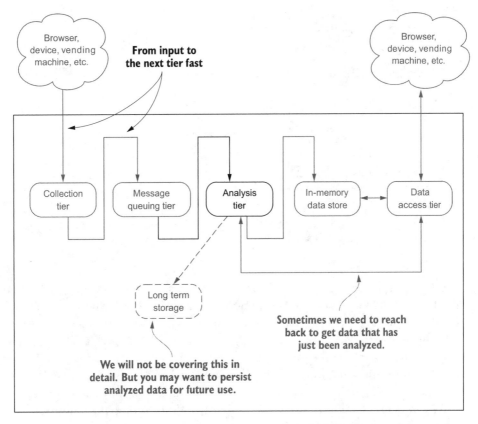

Figure 4.1　The streaming data architecture with the analysis tier in focus

used for this tier and you'll be ready to learn how to perform various operations on the data.

Grab a quick coffee refill, and let's get going.

4.1　*Understanding in-flight data analysis*

Key to understanding the features we'll discuss in this chapter is coming to grips with what *in-flight data* means and grasping the concept of *continuous queries*. If *in-flight* makes you think of something moving in the air, that's the right idea. When it comes to data, *in-flight* refers to all the tuples in the system, from the input source (the message queuing tier), to the output, to a client (the next tier). Data is always in motion and never at rest, meaning it's never persisted to durable storage. (Data *at rest* means the data is stored on disk or another storage medium.) Figure 4.2 shows how this plays out in our streaming architecture.

Figure 4.2 should make it clear that our goal in this tier is to *pull* the data from the message queuing tier as fast as possible; ideally the analysis tier should be able to keep up with the rate at which the collection tier is pushing data into the message queuing

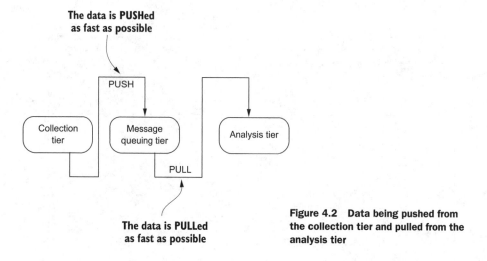

**The data is PUSHed
as fast as possible**

**The data is PULLed
as fast as possible**

**Figure 4.2   Data being pushed from
the collection tier and pulled from the
analysis tier**

tier. How is this different from a non-streaming system? Say, one built with a traditional
DBMS (RDBMS, Hadoop, HBase, Cassandra, and so on)? In those non-streaming systems the data is at rest, and we query it for answers. In a streaming system we turn that
on its head—the data is moved through the query. This model is called the *continuous
query model*, meaning the query is constantly being evaluated as new data arrives.

Let's imagine that you run a large news agency and you want to know if an article is
trending or if a link to it is broken so that you can adjust your marketing campaign or
fix your site. If you were using traditional DBMS technologies, you would have to do
the following:

1 Gather the data from your site
2 Load the data into the DBMS
3 Execute a query to determine if the link is broken or the article is trending
4 Take action
5 Rinse and repeat every *x* minutes or, more likely, hours

Compare that to the steps you might take if you were using a streaming system:

1 Collect the stream of data
2 Start a query that determines if the link is broken or the article is trending
3 Take action

I think you'll agree that it would be hard for your business to react to changes happening now using a traditional system, whereas with the streaming system the query is
always executing against the data, and you can react in real time to trends or problems. In a streaming system a user (or application) registers a query that's executed
every time data arrives or at a predetermined time interval. The result of the query is
then pushed to the client.

Here are the key differences to remember:

- In the architecture of traditional database management systems, when a user (the active party) wants an answer to a question, they submit a query to the system (the passive party), and an answer is returned. This is always based on data that has been loaded into the system before it's queried; in essence, the data set is static.

- In a streaming system a query is started and is continuously (this could be triggered on an interval or another event) executed over the data as it is flowing. The answer to the query is then pushed to the next tier, which may be a user or application.

This inverts the traditional data management model by assuming users to be passive and the data management system to be active. Figure 4.3 shows this inversion graphically.

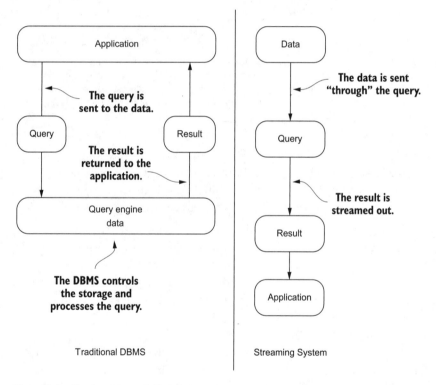

Figure 4.3 Turning things on their head: non-streaming versus streaming

As shown on the left side of figure 4.3, in the traditional DBMS the query is sent to the data and executed, and the result is returned to the application. In a streaming system, as illustrated on the right, this model is completely changed—the data is sent "through" the query, and the result is sent to an application. In the case of the streaming system the data is being pulled or pushed through our system in a never-ending stream; this undoubtedly has implications on both the design and the way we query these systems. To give you a better feel for these differences, table 4.1 highlights some of the main differences between a traditional DBMS and streaming system.

Table 4.1 Comparison of traditional DBMS to streaming system

	DBMS	Streaming system
Query model	Queries are based on a one-time model and a consistent state of the data. In a one-time model, the user executes a query and gets an answer, and the query is forgotten. This is a pull model.	The query is continuously executed based on the data that is flowing into the system. A user registers a query once, and the results are regularly pushed to the client.
Changing data	During down time, the data can't change.	Many stream applications continue to generate data while the streaming analysis tier is down, possibly requiring a catch-up following a crash.
Query state	If the system crashes while a query is being executed, it's forgotten. It's the responsibility of the application (or user) to re-issue the query when the system comes back up.	Registered continuous queries may or may not need to continue where they left off. Many times it's as if they never stopped in the first place.

When you think of all the data zipping around you all day long, from myriad connected devices and appliances to online activity, the questions you could ask and problems you could solve if it all passed through a streaming analysis tier are amazing. Here are some categories and examples to get you going:

- *Tracking behavior*—Imagine being able to provide personalized advertising based on a customer's location, the weather, and their previous buying habits and preferences. McDonald's did this using the VMob platform. In one case study, McDonald's in the Netherlands realized a 700% increase in offer redemptions, and customers using the app returned twice as often and spent on average 47% more (http://mng.bz/1p0t).
- *Improving traffic safety and efficiency*—According to the European Commission (http://ec.europa.eu/transport/themes/urban/urban_mobility/index_en.htm), congestion in the European Union (EU) in and around urban areas costs nearly €100 billion, or 1% of EU GDP annually. According to the Federal Highway Administration (http://ops.fhwa.dot.gov/program_areas/reduce-non-cong.htm), 25% of traffic congestion is caused by traffic incidents. Imagine you were able to employ roadway vehicle sensors (see www.fhwa.dot.gov/policyinformation/pubs/vdstits2007/03.cfm for an introduction); based on our analysis of the traffic data, we can provide drivers with updated traffic conditions and reroute traffic accordingly to maximize driving efficiency. For real-world examples, see Blip Systems

(www.blipsystems.com/traffic/), which has examples of how some cities have solved lots of traffic problems.

- *Real-time fraud analytics*—Every time a credit card is swiped, a complex series of algorithms must be executed to determine whether the attempted transaction is valid or fraudulent. According to FICO (www.fico.com/en/node/8140?file=5582), U.S. fraud losses on credit cards have declined by 70% as a percentage of credit card sales since real-time fraud analytics have been deployed.

These examples are the tip of the iceberg and hopefully have whetted your appetite for what's possible. They also may help you realize that understanding how to build these systems to harness the numerous data streams available in the world today is becoming an essential skill.

But let's not get ahead of ourselves; we have our work cut out for us learning about the core features of an analysis tier. Let's begin our journey by discussing the general architecture of a stream-processing system and then move onto the key features and see how each of the features plays a role in the decision to use a particular framework.

4.2   *Distributed stream-processing architecture*

It may be possible to run an analysis tier on a single computer, but the velocity and volume of the data at some point make this a nonviable option. Instead of tracking trending or broken links to articles, for example, imagine we were interested in analyzing the performance of a gas turbine in real time to determine if it was functioning correctly. According to General Electric, a single turbine engine can produce approximately 1 TB of data per hour. Clearly, using a single computer will quickly become a nonviable option for us. Therefore, we're going to concentrate on the tools and technologies involved in building a distributed analysis tier.

As you survey the technology landscape, you'll find numerous technologies designed for stream processing. At the time of this writing the most popular open source products are Spark Streaming, Storm, Flink, and Samza, all from Apache. I'm not going to go into detail on each of them but will discuss each briefly after going over the generalized streaming architecture so you can see how each fits into it.

A GENERALIZED ARCHITECTURE

Spark, Storm, Flink, and Samza all have the following three common parts:

- A component that your streaming application is submitted to; this is similar to how Hadoop Map Reduce works. Your application is sent to a node in the cluster that executes your application.
- Separate nodes in the cluster execute your streaming algorithms.
- Data sources are the input to the streaming algorithms.

Taking these central ideas and their respective architectures into consideration, we can generalize this into a single common architecture, as shown in figure 4.4.

There are other streaming systems on the market today that have not achieved the same level of popularity as those discussed in the previous section, and undoubtedly

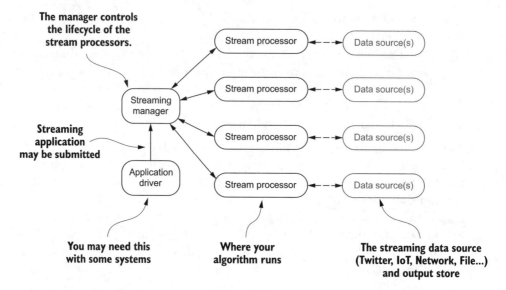

The manager controls the lifecycle of the stream processors.

Streaming application may be submitted

Streaming manager

Application driver

Stream processor   Data source(s)
Stream processor   Data source(s)
Stream processor   Data source(s)
Stream processor   Data source(s)

You may need this with some systems

Where your algorithm runs

The streaming data source (Twitter, IoT, Network, File...) and output store

Figure 4.4   Generic streaming analysis architecture you will find with many products on the market

there will be more in the future. In many cases other products will map onto this common architecture and help in your understanding of how they work. To make sure we're on the same page, consider the common architectural pieces shown in figure 4.4:

- *Application driver*—With some streaming systems, this will be the client code that defines your streaming programming and communicates with the streaming manager. For example, with Spark Streaming your client code is broken into two logical pieces: the driver and the streaming algorithm(s) or job. The driver submits the job to the streaming manager, may collect results at the end, and controls the lifetime of your job.

- *Streaming manager*—The streaming manager has the general responsibility of getting your streaming job to the stream processor(s); in some cases it will control or request the resources required by the stream processors.

- *Stream processor*—This is where the rubber meets the road, the place where your job runs. Although this may take many shapes based on the streaming platform in use, the job remains the same: to execute the job that was submitted.

- *Data source(s)*—This represents the input and potentially the output data from your streaming job. With some platforms your job may be able to ingest data from multiple sources in a single job, whereas others may only allow ingestion from a single source. One thing that may not be obvious from the architectures is where the output of the jobs goes. In some cases you may want to collect the data in your driver; in others you may want to write it out to a different data source to be used by another system or as input for another job.

Now that you know about the various architectures and have boiled them down to our common architecture, let's go back to our example of monitoring the performance of gas turbines to determine if they're functioning correctly and map that to our common architecture. In figure 4.5 you can see our common architecture (simplified to be less busy) with our business problem mapped to it.

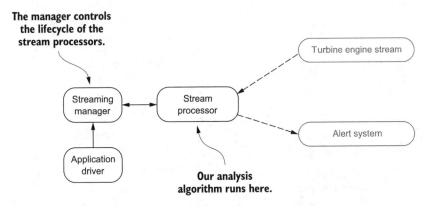

Figure 4.5 Turbine engine monitoring on our common architecture

I realize that this may have been a lot to digest, so take a moment to see if you can map your business problem to the common architecture we've derived. When you're ready, move on to the next discussion of the architecture of the three major streaming systems. Then we'll go a little deeper and talk about some of the key features you'll want to consider when choosing a stream-processing framework.

APACHE SPARK STREAMING

Apache Spark Streaming, often called Spark Streaming, is built on Apache Spark, as depicted in figure 4.6.

Figure 4.6 Apache Spark Streaming with the basic Spark stack

As you'll notice, there are other features built on top of Apache Spark, which is becoming the de facto platform for general-purpose distributed computation. It provides support for multiple languages (Java, Scala, Python, and R) and at the time of this writing has the following high-level tools built on top of it: Spark Streaming, MLlib (machine learning), SparkR (integration with R), and GraphX (for graph processing). Outside the normal project documentation, a great resource to start learning more about Spark is Marko Bonaći and Petar Zečević's book *Spark in Action* (www.manning.com/books/spark-in-action). Figure 4.7 illustrates Spark Streaming's overall architecture.

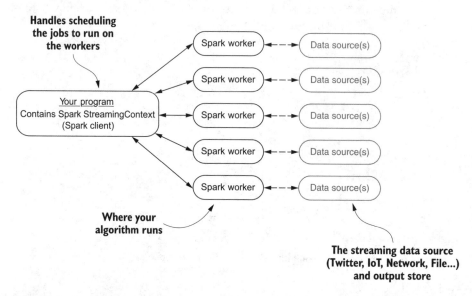

Figure 4.7 Spark Streaming's high-level architecture

Starting from the left in the figure is our program, which contains what is called a Spark StreamingContext; collectively, this is known as *the driver*. Without diving into the details, the Spark StreamingContext contains all the logic to be able to keep track of incoming data, set up the streaming jobs, schedule them on the Spark workers, and execute the jobs. You may notice here that we're talking about jobs and not a stream. Spark and subsequently Spark Streaming operate on batches of work. In the case of Spark Streaming, these batches represent data over a period of time and can be scheduled to run at intervals of less than half a second. A *job* in Spark Streaming is the logic of your program that's bundled and passed to the Spark workers. If you've read about or worked with Hadoop MapReduce, this is the same concept. In the middle of figure 4.7 are the Spark workers, which run on any number of computers (from one to thousands) and are where your job (your streaming algorithm) is executed. As you'll notice, they receive data from an external data source and communicate with the Spark StreamingContext that's running as part of the driver.

APACHE STORM

Apache Storm is tuple-at-a-time stream-processing framework designed for real-time processing of data streams. Storm has so many features I can't cover them in detail. To learn more about Storm, see Sean Allen, Peter Pathirana, and Matthew Jankowski's *Storm Applied* (Manning, 2013). The overall architecture for Storm is shown in figure 4.8.

See
chapter 9
section
9.3.1

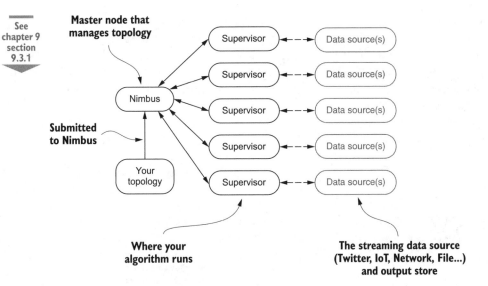

Figure 4.8   High-level overview of Storm architecture showing Nimbus, supervisors, and a data source

You can see that from a high-level Storm is similar to Spark Streaming or perhaps a Hadoop cluster if you change Nimbus to a job tracker and the supervisors to data nodes. Hadoop and Spark use the term *job* to describe the unit of work; with Storm the term is topology. The reasoning behind this is that a *job* will eventually finish, whereas a *topology* will run forever. Let's not get bogged down with semantics; at the end of the day they both represent a way to deploy your program to the worker nodes. With this definition in mind, let's walk through figure 4.8 and discuss the different pieces.

Starting on the bottom left in figure 4.8 you can see the topology is submitted to a component called Nimbus. Nimbus decides how the topology is deployed across the supervisors, assigns different tasks to the supervisors, and monitors the entire system for failures. Moving to the middle of the figure, you see the *supervisor* nodes, where your topology runs. On the right, the *data source* represents the data that will be ingested by the running topology.

APACHE FLINK

Apache Flink is a *stream first* framework, where everything is viewed as a stream. A program written with Flink is composed of two fundamental building blocks: streams and transformations. A *stream* can be considered an intermediate result as the data flows from a source to a sink. A *transformation* is any operation that takes one or more streams as input and produces one or more streams as output. When an application is composed using these concepts and executed with Flink, it's considered a streaming dataflow. The dataflow begins with the ingestion of data from one or more sources, usually contains one or more transformations, and ends with the data being written out to one or more sinks. Flink applications run in a distributed environment composed of a single master and workers. The high-level architecture for this is shown in figure 4.9. We are merely touching the surface here; to learn more about Apache Flink, I strongly recommend you read Sameer Wadkar and Hari Rajaram's *Flink In Action* (Manning, 2016).

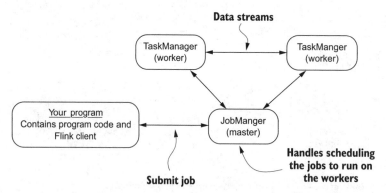

Figure 4.9   High-level Apache Flink architecture

APACHE SAMZA

The streaming model with Apache Samza is slightly different in that it provides a stage-wise stream-processing framework. To do so it uses two prominent technologies found in the Big Data space: Apache Yarn and Apache Kafka. We won't spend much time on those technologies, but I will discuss them briefly as they relate to the high-level Samza architecture shown in figure 4.10.

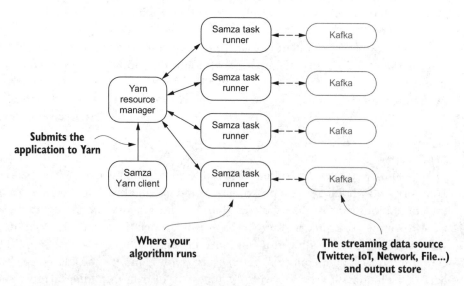

Figure 4.10   High-level Apache Samza architecture

Yarn is a cluster manager designed to handle resource management and job scheduling/monitoring. I know that's a mouthful, but think of it this way: the resource management part is responsible for allocating resources (CPUs, memory, disk, network, and so on) for all applications running on a cluster of computers. The job scheduling/monitoring aspect is responsible for running the job on the cluster. In figure 4.10 you can see that the Samza Yarn client makes a request to the Yarn resource manager asking that the requested resources be allocated for the Samza application to run. Subsequently, after some resource negotiation, the Samza task runners are executed in various nodes in the cluster. This is intentionally simplified because focusing on the Yarn specifics here doesn't add value to our discussion and is subject to change as the project matures. In the center of the figure the Samza tasks are running. In this case all input and output to our Samza tasks will be done using Apache Kafka.

Apache Kafka is a technology that squarely fits into chapter 3's discussion on the message queuing tier and is one we'll revisit in future chapters. For now think of it as a high-speed data store that our streaming tasks will read from and write to. Great resources to learn more about Yarn are Alex Holmes's *Hadoop in Practice*, Second Edition (Manning, 2014) and Chuck Lam, Mark Davis, and Ajit Gaddam's book *Hadoop in Action*, Second Edition (Manning, 2014). For the latest information on Apache Samza, visit http://samza.apache.org.

4.3 Key features of stream-processing frameworks

You can use many different stream-processing frameworks in the analysis tier of our streaming data architecture. We want to pay special attention to a handful of key features when comparing them and deciding whether they're suitable for solving our business problem. The goal is to be able to apply this knowledge when selecting the stream-processing framework you'll use in your streaming data architecture.

4.3.1 Message delivery semantics

In chapter 3 you learned about message delivery semantics with respect to the message queuing tier and the producers, brokers, and consumers. This time the discussion focuses on message delivery semantics on the analysis tier. The definitions don't change, but the implications are a little different.

First, let's refresh your memory on the definitions of the different guarantees:

- *At-most-once*—A message may get lost, but it will never be processed a second time.
- *At-least-once*—A message will never be lost, but it may be processed more than once.
- *Exactly-once*—A message is never lost and will be processed only once.

Those definitions are a slightly more generic version of what you saw before with the message queuing tier. In chapter 3, our focus was on understanding what each of these meant with respect to producing and consuming messages. In this chapter, with the stream-processing tools, we're talking about the continuation of the consumer side of the message queuing tier. How do these manifest themselves in the stream-processing tools you may use in this tier? Let's overlay them on a data flow diagram and walk through them to see what they mean.

Figure 4.11 shows at-most-once semantics with the two failure scenarios: a message dropping and a streaming task processor failing, the latter of which will also result in message loss until a replacement processor comes online.

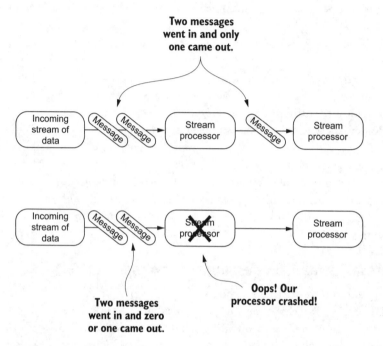

Figure 4.11   At-least-once message delivery shown with the streaming data flow

At-most-once is the simplest delivery guarantee a system can offer; no special logic is required anywhere. In essence, if a message gets dropped, a stream processor crashes, or the machine that a stream processor is running on fails, the message is lost.

At-least-once increases the complexity because the streaming system must keep track of every message that was sent to the stream processor and an acknowledgment that it was received. If the streaming manager determines that the message wasn't processed (perhaps it was lost or the stream processor didn't respond within a given time boundary), then it will be re-sent. Keep in mind that at this level of messaging guarantee, your streaming job may be sent the same message multiple times. Therefore your streaming job must be *idempotent*, meaning that every time your streaming job receives the same message, it produces the same result. If you remember this when designing your streaming jobs, you'll be able to handle the duplicate-messages situation.

Exactly-once semantics ratchets up the complexity a little more for the stream-processing framework. Besides the bookkeeping that it must keep for all messages that have been sent, now it must also detect and ignore duplicates. With this level of guarantee your streaming job no longer has to worry about dealing with duplicate messages—it only has to make sure it responds with a success or failure after a message is processed. Even though being idempotent with this level of messaging guarantee isn't required of your streaming job, I highly recommend that you approach all your streaming jobs with the expectation that they should be idempotent. It will make troubleshooting and reasoning about them much easier.

You may be wondering which of these guarantees you need; it depends on the business problem you're trying to solve. Take our example from earlier—the turbine engine monitoring system. Remember, we want to constantly analyze how our turbine engine is performing so we can predict when a failure may occur and preemptively perform maintenance. Earlier I said our turbines produce approximately 1 TB of data every hour—keep in mind that's one turbine, and we're monitoring thousands to be able to predict when a failure may occur. Do we need to ensure we don't lose a single message? We may, but it would be worth investigating whether our prediction algorithm needs all the data. If it can perform adequately with data missing, then I'd choose the least complex guarantee first and work from there.

What if instead your business problem involved making a financial transaction based on a streaming query? Perhaps you operate an ad network and you provide real-time billing to your clients. In this case you'd want to ensure that the streaming system you choose provides exactly-once semantics.

I hope you're getting the hang of it and can apply this to your business problem. Now let's move on to talk about state management.

STATE MANAGEMENT

Once your streaming analysis algorithm becomes more complicated than using the current message without dependencies on any previous messages and/or external data, you'll need to maintain state and will likely need the state management services provided by your framework of choice. We'll take a simple example to help you understand where and perhaps how state needs to be managed.

Pretend you're the marketing manager for a large e-commerce site and you want to know the number of page views per hour for each visitor. I know you're thinking, "An hour—that can be done in a batch process." Instead of worrying about that right now let's focus on the implied state you must keep to satisfy this business question. Figure 4.12 shows how your streaming task processors would be organized to answer this question.

In figure 4.12 it should be obvious where you need to keep state—right there in the stream processor where your job performs the counting by user ID. If your streaming analysis tool of choice doesn't provide state management capabilities, one viable option for you is to keep the data in memory and flush it every hour. This would work as long as you're okay with the potential of losing all the data if the streaming processor or job fails at any time. As luck would have it, your job would be running smoothly and then, one day, start to fail at 59 minutes into the hour. Depending on your busi-

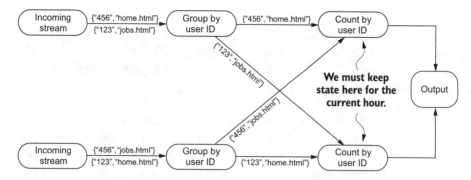

Figure 4.12   Simple example of counting page views per user over an hour

ness case, the risk and possible loss of data by keeping all state in memory may be acceptable. But in many business cases life is not so simple, and you do need to worry about managing state. To help in these scenarios, many stream-processing frameworks provide state management features you can use.

The state management facilities provided by various systems naturally fall along a complexity continuum, as shown in figure 4.13.

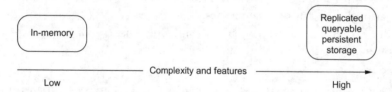

Figure 4.13   State management complexity continuum for stream-processing tools

The continuum starts on the left with a naïve in-memory-only choice, similar to what we used earlier, and progresses to the other end of the spectrum with systems that provide a queryable persistent state that's replicated. If you're saying, "These seem like two totally different slants on state management," you're not alone. The solutions on the low-complexity side only solve the problem of maintaining the state of a computation in the face of failures. For the simple operations of keeping a running count current and not losing track of the current value in the face of failure, these systems are a great fit.

On the other end of the spectrum, the frameworks that offer state management by way of a replicated queryable persistent store help you answer much different and more complicated questions. With these frameworks you can join together different streams of data. Imagine you were running an ad-serving business and you wanted to track two things: the ad impression and the ad click. It's reasonable that the collection of this data would result in two streams of data, one for ad impressions and one for ad clicks. Figure 4.14 shows how these streams and your streaming job would be set up for handling this.

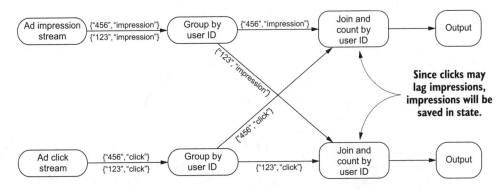

Figure 4.14   Handling ad impression and ad click streams that use stream state

In this example the ad impressions and ad clicks arrive in two separate streams; because ad clicks will lag ad impressions, we'll join the two streams and then count by user ID. Because of the lag in the ad click stream, using a stream-processing framework that persists the state in a replicated queryable data store enables us to join the two streams and produce a single result. I think you'll agree that being able to join streams by using the state management facilities of a stream-processing framework is quite a bit different than making sure the current value of an aggregation is persisted.

If you give additional thought to this example, I'm sure you can come up with other ideas of how to join more than one stream of data. It's a fascinating topic and something we'll look at in more depth in chapter 5. For now, let's continue to the next feature you need to understand when choosing a stream-processing framework.

FAULT TOLERANCE

It's nice to think of a world where things don't fail, but in reality it's not a matter of *if* things will fail, but *when*. A stream-processing framework's ability to keep going in the face of failures is a direct result of its fault-tolerance capabilities. When you consider all the pieces involved in stream processing, there are quite a few places where it can fail. Let's use figure 4.15 to identify all the failure points.

In figure 4.15 are seven points of failure in a simple stream-processing data flow. Go through them and make sure you understand what you'll need from a stream-processing framework with respect to fault tolerance:

1 *Incoming stream of data*—In all fairness, the message queuing tier won't be under the control of the stream-processing framework, but there is the potential for the message queuing system to fail, in which case the stream-processing framework must respond gracefully and not fail if data or the resource isn't available.

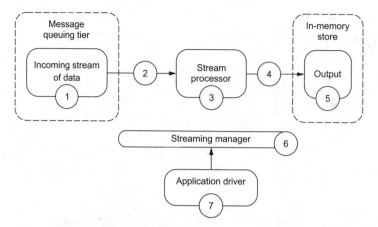

Figure 4.15 The points of failure with stream processing in the context of the streaming architecture

2 *Network carrying input stream*—This is something that the stream-processing framework can't control, but it needs to handle the disruption gracefully.

3 *Stream processor*—This is where your code is running; it should be under supervision of the stream-processing framework. If something goes wrong here—perhaps your software fails or the machine it's running on fails—then the streaming manager should take steps to restart the processor or move the processing to a different machine.

4 *Connection to output destination*—The stream task manager may not be able to control the network path to the output, but it should be able to control the flow of data from the last stream processor so it doesn't become overwhelmed by network backpressure or fail if the network or destination is unavailable.

5 *Output destination*—This wouldn't be under the direct supervision of the stream task manager, but its failing could impact the processing of the stream, so it needs to be considered.

6 *Streaming manager*—If this fails, you end up with a situation called *running headless*—the stream processors would continue to run without being supervised by the streaming manager. If this component fails, there's no supervisor for the data flow and the stream processors—no new ones can be started or failed ones recovered.

7 *Application driver*—This comes in two flavors. With the first, the application driver does nothing more than submit the streaming job to the streaming manager—we're not worried about this type. The second flavor is where the application driver logically contains the streaming manager and in turn is subject to the same risk as the streaming manager.

Let's go through how these problems are solved or could be solved. First, let's boil our problem down a bit. In the preceding list, we can eliminate the incoming stream and output destination availability from the concerns of the streaming framework. It should go without saying that the streaming framework must not fail if there are failures with the input or output destinations. For this discussion we won't consider those aspects to be fault tolerance–related. If we take the list and consolidate it down to the common elements, we end up with the following:

- *Data loss*—This covers data lost on the network and also the stream processor or your job crashing and losing data that was in memory during the crash.
- *Loss of resource management*—This covers the streaming manager and your application driver, in the event you have one.

In the discussion of fault tolerance in chapter 3, I talk about ways to prevent data loss. When it comes to stream-processing frameworks, all the common techniques for dealing with failures involve some variant of replication and coordination. A common approach would be for the stream manager to replicate the state of a computation (the state of your streaming job) onto different stream processors. If there's a failure, then the streaming manager must coordinate the replicas in order to recover properly from failures. It's common for fault-tolerance techniques to be designed with a tolerance up to a predefined number of simultaneous failures, in which case you'll hear of a system being called *k-fault tolerant*, where *k* represents the number of simultaneous failures.

In general there are two common approaches—state-machine and rollback recovery—a streaming system may take toward replication and coordination. Both assume that we've designed and thought about our streaming algorithm in an idempotent way. If you recall from before, for our streaming job to be idempotent it means that two non-faulty stream processors that receive the same input in the same order will produce the same output in the same order. If we can ensure that, then we can refer to those two stream processors, and hence our streaming job, as idempotent and consistent if they generate the same output in the same order.

The first approach used by stream-processing systems is known as *state-machine*. With this approach the stream manager replicates the streaming job on independent nodes and coordinates the replicas by sending the same input in the same order to all. This approach requires *k* + 1 times the resources of a single replica, where *k* is the number of stream processors our streaming job is running on. But this allows for quick failover, resulting in little disruption. For some applications, such as an intrusion-detection system that has low-latency requirements at all times, the extra resource cost may be justifiable.

The second approach is known as *rollback recovery*. In this approach the stream processor periodically packages the state of our computation into what is called a *checkpoint*, which it copies to a different stream processor node or a nonvolatile location such as a disk. Between checkpoints, the stream processor has to keep track of the computation. Given the relative high latency of disks, once they're introduced the latency of our streaming computation will go up. It therefore may not be unreasonable for a stream-processing framework to instead decide to take the approach of copying the checkpointed state to other stream processor nodes and also maintain logs in memory. In this case, if a stream processor fails, the stream manager would need to reconstruct the state from the most recent checkpoint and replay the log to recover the exact pre-failure state of the streaming job. Compared to the first approach, this approach has a lower overhead, but it's more expensive in terms of time to recover when a failure does happen. This approach is useful in situations where fault tolerance is important and rare moderate latencies are acceptable.

As you investigate which stream-processing framework to use to solve your business problem, you'll find that if they offer fault-tolerance, they'll all be some variant on these two common approaches. If you're interested in taking a deeper dive into either of these approaches, you may find the following articles of interest: Elnozahy, Alvisi, Wang, and Johnson's "A Survey of Rollback-Recovery Protocols in Message-Passing Systems"[1] and Schneider's "Implementing Fault-Tolerant Services Using the State Machine Approach: A Tutorial."[2]

4.4   Summary

This chapter dove into the common architecture of stream-processing frameworks that you'll find when surveying the landscape and covered the core features that you need to consider.

You learned about the following:

- The common architecture of stream-processing frameworks
- What message delivery semantics mean for this tier
- What state is and how it can be managed
- What fault tolerance is and why you need it

I understand that some of this may seem fuzzy or fairly abstract; don't worry about it at all. In chapter 5 we'll focus on how to perform analysis and/or query the data flowing through the stream-processing framework. Some may say that's where the fun begins, but to effectively be able to ask questions of the data, you need the understanding you developed in this chapter.

Are you ready to start asking questions of the data? Great! Turn the page and get started.

---

[1]   *ACM Computing Surveys*, 34(3):375–408 (2002, www.cs.utexas.edu/~lorenzo/papers/SurveyFinal.pdf.
[2]   *ACM Computing Surveys*, 22(4):299–319 (1990), www.cs.cornell.edu/fbs/publications/smsurvey.pdf.

Algorithms
for data analysis

This chapter covers

- Querying a stream
- Thinking about time
- Understanding four powerful summarization techniques

Chapter 4 covered how the data flows through many stream-processing frameworks, the delivery semantics, and fault tolerance. In this chapter we're going to depart from the architectural views and discuss the algorithmic side of stream processing, often called *streaming analytics* or *stream mining*. We will focus on the *what* and *why* of streaming analysis algorithms and occasionally dip our toes into the detailed *how*. Don't worry if you're looking for the detailed math or code behind the algorithms—ample resources will be provided so that you can continue your learning.

Before we begin, I'll talk about how we perform queries with these tools. In general, there are two types of queries that you may want to execute in a streaming system:

- *Ad-hoc queries*—These are queries asked one time about a stream. For example: What is the maximum value seen so far in the stream? This style of query is the same kind you would execute against an RDBMS.

- *Continuous queries*—These are queries that are, in essence, asked about the stream at all times. For example: Determine the maximum value ever seen in the stream emitted every five minutes and generate an alert if it exceeds a given threshold.

Unfortunately, in the current technology landscape full of so many different stream-processing frameworks, no two systems offer the same query language, and in many cases there is no SQL-like query language available. Instead you express the algorithmic details programmatically. Table 5.1 shows the current state of query language support in each of the popular stream-processing frameworks (subject to change, as many of these projects are being actively developed and are all maturing).

Table 5.1   Stream-processing framework query language support

Product	Query language support
Apache Storm	As of version 1.1.0 Apache Storm has had SQL support (http://storm.apache.org/releases/1.1.0/storm-sql.html). As of this writing it is still considered experimental and not ready for production use
Apache Samza	Since version 0.9 of Apache Samza there has been a JIRA open for adding query language support. As of this writing, that JIRA is still open, and Samza does not have any query language support: https://issues.apache.org/jira/browse/SAMZA-390.
Apache Flink	Table API supporting SQL-like expressions (http://ci.apache.org/projects/flink/flink-docs-release-0.9/libs/table.html).
Apache Spark Streaming	SparkSQL/Hive language support (http://spark.apache.org/docs/latest/sql-programming-guide.html).

Given the current state of SQL-like support in the market today, I won't show implementation details for each product because they're all different. But I will provide guidance on implementing each algorithm with each stream-processing framework. With a high-level understanding of the general way we may have to perform different stream-mining activities, let's discuss the constraints we must keep in mind.

5.1   *Accepting constraints and relaxing*

As you know from previous chapters, one of the unique aspects of a streaming system is that we can't store the entire stream because it's unbounded and never-ending. Our goal is to continually provide results to queries online. As data reaches the analysis tier, the results must be recomputed or updated and potentially emitted. On the surface, answering these types of queries may seem easy, but when you consider or design algorithms that will process a stream, it is important to take into consideration the following constraints:

- *One-pass*—You must assume that the data is not being archived and that you only have one chance to process it. This can have significant consequences on your algorithmic development. For example, many traditional data-mining algorithms are iterative and require multiple passes over the data. To work in a streaming scenario, each of these needs to be modified accordingly. I find it helpful to remember that you only get to touch the data one time.

- *Concept drift*—This is a phenomenon that may impact your predictive models. Concept drift may happen over time as your data evolves and various statistical properties of it change. Depending on the type of analysis you are doing and the predictive models you have developed, you may need to take this into consideration.

- *Resource constraints*—For many data streams we have little to no control over the arrival rate of the data. There may be times when, due to a temporary peak in the data speed or volume and the resources at our disposal, an algorithm may have to drop tuples that can't be processed in time, called *load shedding*. This constraint is almost universal in streaming systems, but surprisingly few algorithms take it into account. There are two general types, random and semantic; the latter makes use of properties of the stream and quality-of-service parameters.[1]

- *Domain constraints*—Whereas the other constraints are almost universal to all data streams, these are particular to your business domain. For example, if our social network had 100,000,000 users and we wanted to do an analysis of all emails sent between users, we would need to be able to store double that amount of email addresses. Our storage requirements are easily in the multiple-petabyte range. Being able to do simple statistics or distinct counts about this stream would be challenging. This may appear to be a resource constraint, but it's our business data that causes the constraint.

It is because of these constraints that virtually every streaming method uses some form of synopsis. The basic idea we will see employed is an online synopsis that is used for analysis. Many different kinds of synopsis can be created; as you will see, the exact kind used will have a strong influence on the type of questions that can be answered. Before we dig into these different mining activities, let's look at time as it relates to stream processing and its impact on streaming analysis.

5.2 Thinking about time

If you've worked with a data system where the data is static, such as Hadoop or an RDBMS, you probably thought about time as you were executing queries. In a static world you execute your MapReduce job, Spark job, Hive query, SQL query, or in some other fashion query the data set and perhaps provide a time range in the where clause, and you know the resulting data is all the data that is loaded within a given time range. In contrast, with a streaming system, along with our constraints, the data is constantly flowing. It may be out of order when we see it or delayed—and we can't query all the

---

[1] For more information, see "Load Shedding in a Data Stream Manager" in *Proceedings of the 29th International Conference on Very Large Data Bases* (2003, pages 309–320), http://dl.acm.org/citation.cfm?id=1315479.

data at once, because the stream never ends. Don't worry—all is not lost. I'll discuss concepts and approaches to thinking about time and solving common problems when analyzing a stream of data.

STREAM TIME VS. EVENT TIME

Stream time is the time at which an event enters the streaming system. *Event* time is the time at which the event occurs. Imagine we are collecting data from a fitness-tracking device such as a Fitbit, and the data is flowing into our streaming system. Stream time would be when the fitness event enters the analysis tier; event time would be when it takes place on the device. Thinking back to our overall architecture, stream time is when the event first enters the analysis tier. If the streaming analysis you're doing relies on event time, realize that it's often not the same as stream time. Often there will be a variance, called *time skew*, sometimes significant, between when an event is created and when it enters the system, as shown in figure 5.1.

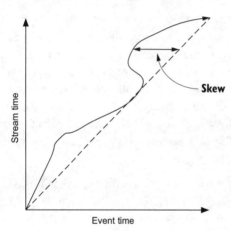

Figure 5.1 Time skew between event time and stream time

Taking into consideration our working example, how would this impact our analysis of the data? How will the drift impact the average speed for the runners we're tracking? Our ability to answer these questions is directly related to the next topic: windowing techniques found in stream-processing systems. Keep the concept of time skew in mind, and we will come back to these questions.

WINDOWS OF TIME

Due to its size and never-ending nature, the stream processing engine can't keep an entire stream of data in memory. This means we can't perform traditional batch processing on it. How then do we perform computations on it? The answer is: by using windows of data. A *window* of data represents a certain amount of data that we can perform computations on. Figure 5.2 shows that a window of data is a small amount of the data flowing through the system at a given point in time.

In figure 5.2, you see that the window is indeed a small part of the entire stream of data. It is a little more complex than that, but not much. The added complexity comes

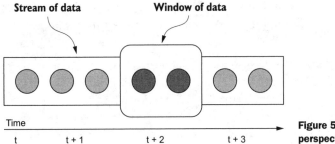

Figure 5.2 A window of data in perspective to the rest of the stream

by the way of two attributes common to all windowing techniques: the trigger and eviction policies. The *trigger* policy defines the rules a stream-processing system uses to notify our code that it's time to process all the data that is in the window. The *eviction* policy defines the rules used to decide if a data element should be evicted from the window. Both polices are driven by either time or the quantity of data in the window. The distinction between the two policies and how time or the count of items come into play will become clearer as we discuss windowing techniques, of which the two most prominent in practice are sliding and tumbling.

5.2.1 Sliding window

The *sliding* window technique uses eviction and trigger policies that are based on time. The two policies are manifested in the window length and sliding interval, as shown in figure 5.3.

The window length represents the eviction policy—the duration of time that data is retained and available for processing. In figure 5.3 the window length is two seconds; as new data arrives, data that is older than two seconds will be evicted. The sliding interval defines the trigger policy. In figure 5.3, the sliding interval is one second.

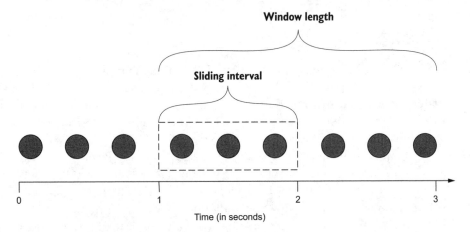

Figure 5.3 Sliding window showing the slide interval and the window length

This means that every second our code would be triggered, and we would be able to process the data in the sliding interval as well as the entire window length.

EXAMPLE USAGE

Going back to our Fitbit example, remember that we have the data flowing into our streaming system. The head of product marketing has asked us to build a dashboard that shows the average speed for all runners broken down by age groups, such as 12–17, 18–24, 25–34, and so on. The dashboard should be updated every 5 seconds, and the averages should represent data for the last 30 minutes. Don't worry about the dashboard aspect; concentrate on the streaming analysis. How you would you handle this using the sliding window technique?

We would want a window length of 30 minutes and a sliding interval of 5 seconds. Remember to take into consideration stream time versus event time. Will your analysis make sense if the window length and sliding interval are based on stream time?

FRAMEWORK SUPPORT

Not all current stream-processing frameworks support sliding windows or provide the same level of support. Table 5.2 identifies the level of support for sliding windows in each of the popular frameworks.

Table 5.2 Sliding window support in popular stream-processing frameworks

Framework	Sliding window	Event or stream time	Comments
Spark Streaming	Yes	Stream time	Spark Streaming doesn't allow custom policies.
Storm	No	N/A	Storm doesn't provide native support for sliding windowing, but it could be implemented using timers.
Flink	Yes	Both	Flink allows a user to define a custom policy and trigger policies.
Samza	No	N/A	Samza doesn't provide direct support for sliding windows.

The details of windowing support for Spark Streaming and Flink are both well documented on their respective project sites. Note that Spark Streaming only supports windowing using stream time. If your application is sensitive to the differences between stream time and event time, you will need to make sure your windowing sizes and algorithms account for this.

For both Apache Storm and Apache Samza, it may be possible to implement sliding window support, but it's not natively supported by either of those tools. So, the work you would have to do may be substantial and not as efficient as a framework that natively supports sliding windows. Delving into the details of implementing this support in either framework is beyond the scope of this text. If that's something you need, check the latest additions of each as well as their JIRA tickets and email lists for discussions on windowing support. Considering that they're all open source projects, you may also contribute enhancements to one of the projects.

5.2.2 Tumbling window

A tumbling window offers a slight twist on the windowing concept. The eviction policy is always based on the window being full, the trigger policy is based on either the count of items in the window or time, and they break down into two distinct types: count-based and temporal-based. First let's consider count-based tumbling; figure 5.4 shows how this works.

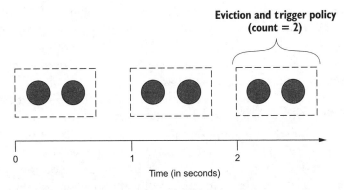

Figure 5.4 Count-based tumbling window with an eviction and trigger policy of two

In figure 5.4 both the eviction and trigger policies are equal to two: when two items are in the window, the trigger will fire, and the window will be drained. This behavior is irrespective of time—whether it takes one second or five hours for the window to fill, the trigger and eviction polices will still execute when the count is reached.

Compare that to the temporal tumbling window in figure 5.5, a tumbling window with an eviction and trigger policy of two seconds.

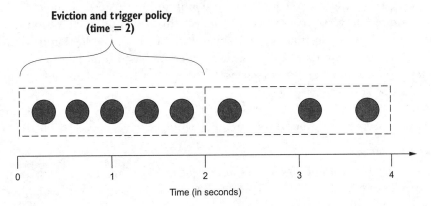

Figure 5.5 Temporal tumbling window with an eviction and trigger policy of two seconds

In the case of figure 5.5, both policies are based on a two-second time frame. In this case it doesn't matter if there are three tuples or five tuples in the window. When the time lapses, the trigger and eviction policies will fire, and the window will be drained. This is distinctly different from the sliding window described in the preceding section.

EXAMPLE USE

Let's imagine that we manufacture a bicycle that is equipped with various sensors, which emit data points such as GPS coordinates, current speed, current direction, ambient temperature, and humidity. From this data set we are interested in understanding two metrics. First, we want to know the average speed of all our bikes every 30 seconds throughout the day. We may break this down by geography, but for now we want a global count. Second, we want to know every time there are more than 100 people riding one of our bikes in a city. Take a moment and jot down how you would handle these two scenarios using tumbling windows.

How did you do? For the first metric, we want our code to be triggered every 30 seconds. To ensure this, I would create a stream that contains only the speed measurement from our sensors and set up a temporal tumbling window of 30 seconds. When our code is triggered, we compute the average using all the tuples in the window at that time. To break this down by geography later, we have a couple of options. One way would be to not pre-filter the stream to contain only the speed measurement, but have it contain the full message sent from the bicycle. Then every 30 seconds we can extract the speed and we would also have the GPS coordinates in hand that we could use to segment the data by any geographic boundary we wanted.

A second way would be to do more filtering. Taking this approach, we would have our collection tier split the data out by geography first and then send it through the rest of the tiers. Then we would have a specific stream: speed with geography. This could be problematic and fairly inflexible. We would need to determine ahead of time the geographic boundaries we used for segmentation and have a strategy for how to handle changes to them.

Let's now consider the second metric we want to capture: every time there are 100 people riding our bicycles in a city. To support this we would need to do two things. First, we create a stream (that may or may not start from our collection tier) for every new city we see in the data and then set up a count-based tumbling window using a window size of 100. When the trigger policy executes, we would have all the tuples for each city that reached 100 cyclists.

Okay, we've worked through two fairly simple examples. Now let's take a look at the current framework support for tumbling windows.

FRAMEWORK SUPPORT

Not all current stream-processing frameworks support tumbling windows or provide the same level of support. Table 5.3 shows the level of support for tumbling windows in each of the popular frameworks.

Table 5.3 Tumbling window support in popular stream-processing frameworks

Framework	Count	Temporal	Comments
Spark Streaming	No	No	Currently you would need to build this.
Storm	Yes	Yes	Although Storm does not have the native windowing support, we can easily implement this.
Flink	Yes	Yes	Flink has built-in support for both types of tumbling windows.
Samza	No	Yes	Samza does not provide direct support for sliding windows.

At the time of this writing Apache Flink is the only framework that has built-in support for tumbling windows, both count- and temporal-based. For the other frameworks the level of effort to implement tumbling window support varies. As with all software, the features available when you evaluate it will likely have changed, so if you need tumbling windows to solve your business problem, double-check the feature set of your chosen tool.

We have now taken a look at the two most common types of windowing found in modern stream-processing frameworks. This information is important to keep in mind as we discuss summarization techniques.

5.3   *Summarization techniques*

In this section we are going to explore four summarization techniques that form the basis for many different types of analysis you may perform as well as other data-mining techniques you may use. You may wonder why we need to talk about summarizing a stream and question why we need to settle for non-exact answers to questions. The answer lies in the nature of stream processing. Remember, we don't know if the stream will ever end, nor can the entirety of it fit in memory. That makes it extremely difficult to provide exact answers to questions about the data in the stream. In many cases, having a high degree of confidence that the answer to a question is correct or correct enough is adequate. Admittedly you may run into situations where an exact answer must be known, but providing that level of exactness will come at a cost of processing speed and/or implementation. When you are approached with a request to provide exact numbers, it is important to dig in and find out whether a good estimate would work.

> **NOTE** I once worked on a streaming analytics project where we were told our numbers had to be exact because that is how things had always been done in the past (in the pre-streaming world). But due to how the clients were consuming the data, they could not end up with exact metrics. Do you know what happened? You're right—nothing, because the reality was the picture of the business did not change. As humans, we are good at seeing patterns, and if the data being emitted from a stream-processing application is representative

of the events occurring in a business—but down-sampled so there is less data—the picture will have the same shape when visualized.

Some of the techniques I cover next are a little deeper. Take your time and if you need to, take it slowly, section by section. Ready? Good, let's now dig into our first summarization technique: random sampling.

5.3.1 *Random sampling*

Often you may want to take a random sample from a stream. Pretend that we have built a popular advertising network and our ad servers receive 10 million ad views per minute. That's great, but now we want to perform a statistical analysis of the ad serving as it is happening. On the surface that seems pretty easy, but as you think about it you realize that this data is moving fast, it never stops, and it doesn't fit into memory. A viable solution would be to sample the stream as it is flowing. How do we take a random sample from a data set that you can't hold in memory or on disk? How do we know it's random?

A common approach to solving this problem is to use a technique called reservoir sampling. *Reservoir sampling* is based on the notion that we can hold a predetermined number of stream values (the reservoir), and when a new one arrives we can probabilistically determine whether to add it to our collection or randomly select one of the values already in the reservoir as the random sample. Figure 5.6 shows the general flow of reservoir sampling; as new data arrives it goes through a sampling algorithm, and a random sample is determined.

Let's look at what is happening at each step in figure 5.6. Remember, our goal is to ensure that after we process the 16th item, the elements in the reservoir represent a random sample of all the data we have seen, and we have selected a random value. No

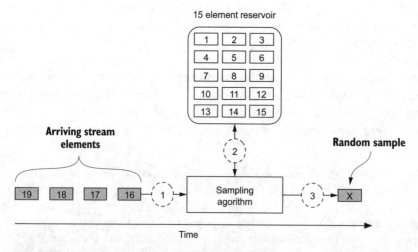

Figure 5.6   General flow of reservoir sampling with first new data item about to be processed

matter how many elements have been consumed from the stream, each element has the same probability of being included in the reservoir. Keep in mind that figure 5.6 shows the state of the reservoir after we have processed the first 15 items. We are using 15, but the general rule is the reservoir is always filled with the first *x* values in the stream, where *x* is the size of the reservoir. After the reservoir is filled and our application is running for a while, we would expect the reservoir to contain a more distributed but random data set.

With that in mind, let's discuss the steps identified in figure 5.6:

1 When the 16th data item arrives, we need to determine if it should be added to the reservoir with a probability of k/n, where k is the size of the reservoir and n is the data element number we are processing. Using these values, the probability that this element should be inserted into the reservoir is 15/16, because we have a reservoir of 15 and we are processing the 16th element.

2 To decide if we add element 16, we generate a random number between 0 and 1. If it is less than 15/16, then we add it to the reservoir and displace one of the items already in the reservoir. If the random number is greater than 15/16, then item 16 becomes our random sample.

3 If element 16 is added in step 2, then we randomly select any element in the reservoir and replace it with the 16th element. The item selected is the random number we use.

That's reservoir sampling. Our next step would be to integrate it into our streaming analysis framework of choice. Currently this algorithm is not provided out of the box with any of the frameworks we have been discussing (Spark Streaming, Storm, Samza, or Flink), but implementing this with any one of them should be fairly straightforward. To learn more about reservoir sampling, the original paper, Jeffrey Vitter's "Random Sampling with a Reservoir" (*Association for Computing Machinery Transactions on Mathematical Software*, 1985, available at www.cs.umd.edu/~samir/498/vitter.pdf), is a great place to start.

5.3.2 *Counting distinct elements*

You may want to count the distinct items in a stream, but remember we are constrained by memory and don't have the luxury of storing the entire stream. In this section we continue with our ad network example from section 5.3.1, where we have an ad network that is serving 10 million ad views per minute. We're going to try and answer this question: How many distinct ads were shown in the last minute?

The preceding section showed how to take a random sample of that data flowing, but if we wanted to count the distinct ads shown every minute, how would we do that? You may be thinking, "It's only 10 million items—I can store that in a hash table or other data structure that provides search capabilities, and the problem is solved." That may be the case for our ad server, but what if we were building a network intrusion detection system that had to operate at 40 Gbps (~78 million packets per second, assuming 64-byte packets)? In that case, and in any case where we can't store the entire stream, we need to rely on probabilistic algorithms to generate our distinct counts.

There are two general categories of algorithms used to solve this problem:

- *Bit-pattern-based*—The algorithms in this class are all based on the observation of patterns of bits that occur at the beginning of the binary value of each element of the stream. Using the bit pattern—more specifically, the leading zeros in the binary representation of a hash of the stream element—the cardinality is determined. Some of the algorithms you would find in this category are LogLog, HyperLogLog, and HyperLogLog++.
- *Order statistics-based*—The algorithms in this class are based on order statistics, such as the smallest values that appears in a stream. MinCount and Bar-Yossef are two algorithms you would find in this category.

In modern practice the bit-pattern algorithms are most commonly used and are the focus of the remainder of this section.[2]

Let's now turn our attention to the bit-pattern-based algorithms; the most popular and prevalent in practice are HyperLogLog and HyperLogLog++. Conceptually, HyperLogLog and HyperLogLog++ are the same, so I will refer to them collectively as HyperLogLog for this discussion. Figure 5.7 shows the general flow of the algorithm.

Figure 5.7 shows the general flow of processing a new element with the HyperLogLog algorithm. Let's walk through it from the top.

- In step 1 is the ad ID that was viewed. In this case I've used a UUID—there's nothing special about using a UUID; for your data, if you have IDs, you could use them.
- In step 2 the string from step 1 is passed through a hash function, resulting in the hashed value you see before step 3.
- Step 4 is where the magic begins. Here we take the binary string of the hashed value from step 3 and determine which register value, often called the bin, to update and the value to update it with. The six least significant bits are used to determine which register value position will be updated. The number of bits used is called the *precision*; I chose six arbitrarily. If you use this algorithm for your analysis, make sure you understand the precision implications. The binary value of those bits 100010 is 34. Therefore, we are going to be updating the value at index 34.

---

[2] To learn more about the order statistics–based algorithms, a couple of good jumping off points are Ziv Bar-Yossef's "Counting Distinct Elements in a Data Stream" (*Randomization and Approximation Techniques*, 2002) at https://link.springer.com/chapter/10.1007/3-540-45726-7_1, and Frederic Giroire's "Order Statistics and Estimating Cardinalities of Massive Data Sets" (*International Conference on Analysis of Algorithms*, 2005) at www.emis.ams.org/journals/DMTCS/pdfpapers/dmAD0115.pdf.

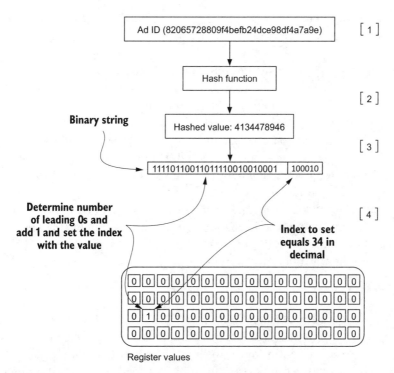

Figure 5.7   Processing a single stream element with the HyperLogLog algorithm

- Now that we know the index that will be updated, we determine the number of leading zeros, starting from the right, for the rest of the bit string and add 1 to it. In this case there are no 0s, so we end up with 0 + 1, and we update index position 34 with the value of 1.
- At this point you can determine the distinct counts (again, it's an approximation) by taking the harmonic mean of all the register values.

That is the general flow of the algorithm. With this algorithm keep in mind that the count of leading zeros in a bit string is used to estimate the cardinality of a stream. Then to increase accuracy, the average of many estimates is taken to reduce bias and the harmonic mean is used to reduce the impact of outliers. These algorithms have their start with, and are enhancements to, the original work by Philippe Flajolet and G. Nigel Martin's "Probabilistic Counting Algorithms" (*Journal of Computer and Systems Science*, 1985) and more recently Durand and Flajolet's "LogLog Counting of Large Cardinalities" (*Annual European Symposium on Algorithms*, 2003).

HyperLogLog++ provides several improvements over HyperLogLog, namely in the reduction of memory usage and an increase in accuracy for a range of cardinalities. Our focus has been on how these algorithms work conceptually so you know how to think about and use them.[3] In practice this algorithm isn't hard to implement, and in fact you may be able to find implementations readily available in the language you're using.

A couple of other things to keep in mind regarding HyperLogLog are that it uses little space and is distributable. From a size and space standpoint, according to the authors of the papers I mentioned, you can count one billion distinct items with an accuracy of 2% using only 1.5 K of memory, which is quite impressive. From the distributed standpoint it is easy to perform a union operation between two HyperLogLog structures. When doing stream analysis, this will enable you to maintain summarizations on each node that is analyzing data and then join them to determine an overall approximate, distinct count.

You should also be able to integrate this into any of the streaming frameworks we've been looking at. With this information you can now determine the approximate distinct counts for your stream. In the next section we will look at an algorithm that helps us answer a slightly different question.

5.3.3 *Frequency*

The preceding section discussed determining the distinct count for a stream. In this section we'll try to answer this question: How many times has stream element X occurred?

See chapter 9 section 9.3.2 and code listing 9.25

The most popular algorithm for answering this type of question is called Count-Min Sketch.[4] This algorithm can be used any time you need count-based summaries of your data stream. In general Count-Min Sketch is designed to provide approximate answers to the following types of questions:

- A *point query*—You are interested in a particular stream element.
- A *range query*—You are interested in frequencies in a given range.
- An *inner product query*—You are interested in the join size of two sketches. For our ad example we may use this to provide a summarization to this question: What products were viewed after an ad was served?

These three types of questions are fundamental to a lot of streaming applications. In our ad-serving example, we may want to ask how often ad X has been viewed. You will also find that similar questions are fundamental to network monitoring and analysis, where millions of packets per second are processed and there is a strong desire to prevent malicious intent such as a Denial Of Service (DOS)

---

[3] To understand the inner workings of these algorithms I encourage you to read Flajolet, Fusy, Gandouet, and Meunier's "HyperLogLog: The Analysis of a Near-optimal Cardinality Estimation Algorithm" (*Conference on Analysis of Algorithms,* 2007) at http://algo.inria.fr/flajolet/Publications/FlFuGaMe07.pdf and Huele, Nunkesser, and Hall's "HyperLogLog in Practice: Algorithmic Engineering of a State of the Art Cardinality Estimation Algorithm" (*Proceedings of the EDBT,* 2013 Conference) at https://static.googleusercontent.com/media/research.google.com/en//pubs/archive/40671.pdf.

[4] Graham Cormode and S. Muthu Muthukrishnan first published an article on this algorithm in the *Journal of Algorithm* (2004) titled "An Improved Data Stream Summary: The Count-Min Sketch and Its Applications." You can read it at http://dimacs.rutgers.edu/~graham/pubs/papers/cm-full.pdf.

attack.[5] I'm sure you can come up with many more examples of when the Count-Min Sketch algorithm could be useful; for now let's dig into how this works.

Count-Min Sketch, as its name implies, was designed to count first and compute the minimum next. Let's get a of couple of definitions out of the way before we see how this works diagrammatically. Count-Min Sketch is composed of a set of numeric arrays, often called *counters*, the number of which is defined by the width w and the length of each is defined by the length m. Each array is indexed starting at 0 and has a range of $\{0...m - 1\}$. Each counter must be associated with a different hash function, which must be pairwise independent—otherwise the algorithm won't work as designed. How this all comes together is shown in figure 5.8.

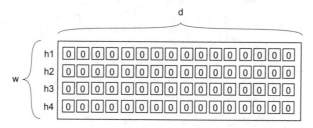

Figure 5.8   Setup of the Count-Min Sketch algorithm

As figure 5.8 shows, this is a 2-dimensional array with all elements initialized to 0 and each row associated with a different hash function. Using different hash functions increases the accuracy of the summary while also reducing the probability of bad estimates, as the chance of hash collisions has been reduced. For our ad network example, the sketch will represent a probabilistic summarization of how many times an ad was served. If our sketch looked like figure 5.8, we would have a 4 x 16 2-dimensional array. Each row is independent and represents a bit array that we'll use to keep count.

Now let's walk through the process of updating the sketch as ad view data is streaming into our system, as shown in figure 5.9.

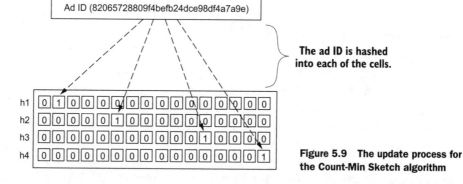

The ad ID is hashed into each of the cells.

Figure 5.9   The update process for the Count-Min Sketch algorithm

---

[5] For an idea of how this type of algorithm is used in network monitoring and analysis, a good place to start is with Cormode and Muthukrishnan's "What's New: Finding Significant Differences in Network Data Streams (INFOCOM, 2004) at http://infocom2004.ieee-infocom.org/Papers/33_1.PDF.

In figure 5.9 we have an ID of an ad that we want to add to the sketch. The first step is to hash the value using the hash function for each respective row and then increment the count for the cell the value hashes to by 1. For our example, all the values were 0, so the result counts are all 1. As ad view data continues to flow through our streaming system, we will repeat this process of updating our sketch. After some time has passed, we want to estimate how many times our ad from figure 5.9 was viewed. To get this frequency estimate, we would use the following equation:

ESTIMATED COUNT =
min(h1(82065728809f4befb24dce98df4a7a9e),h2(82065728809f4befb24dce98df4a7a9e),
h3(82065728809f4befb24dce98df4a7a9e),h4(82065728809f4befb24dce98df4a7a9e))

In that function we're hashing the ID of the ad we're interested in. That gives us the four cells to look at. That is a salient point that may not be obvious from the example equation.

Specifically the result of h1 is the hash that determines the counter to look up. This is the same for h2, h3, and h4. We then take the minimum value from the four cells. This value represents the approximate count for the number of times the ad was viewed. Keep in mind that this algorithm will never undercount, but could overcount. How accurate is this? In the original paper, the authors show that with a width of 8 and a count of 128 (a 2-dimensional array of 8 x 128) the relative error was approximately 1.5%, and the probability of the relative error being 1.5% is 99.6%.

I find it fascinating that we can do this with little space and with little computational cost. This is a pretty straightforward algorithm that can be used to answer a lot of questions. To learn more and gain a deeper understanding of the why behind it, read the award-winning paper by Cormode and Muthukrishnan, "An Improved Data Stream Summary: The Count-Min Sketch and Its Applications (*Journal of Algorithm*, 2004).[6]

Up next we're going to talk about a sketch that is closely related to the Count-Min Sketch, except this one is used when you want to determine whether you've seen a stream element before.

5.3.4 *Membership*

The question we're asking now is: Has this stream element ever occurred in the stream before?

That may seem like a tall order to fill. We know from earlier discussions that we can't store the whole stream—realistically we can't even store an ID for every element we've seen in the stream. You may wonder how then are we going to pull this off? Simple. We're going to use a data structure that you should look to when trying to answer membership type queries: a *Bloom filter*. A Bloom filter is tailor made for this specific task. As with the other algorithms we've seen in this chapter, the Bloom filter's accuracy is probabilistically bound, and as expected, this is configurable.

---

[6] The paper can be found at http://dimacs.rutgers.edu/~graham/pubs/papers/cm-full.pdf.

A unique feature of Bloom filters is that false positives are possible, but false negatives are not. What exactly does that mean? It means that if the filter reports that the stream element has not been seen, then that will always be true. But if the filter reports that the element *has* been seen before, then that may or may not be true. In the literature there are various advanced Bloom filters, but for our discussion we'll stick to the good old plain one. Once you understand how it works, you'll be ready to take on more complex ones.

A Bloom filter is composed of a binary bit array of length m and is associated with a set of independent hash functions. Does that sound familiar? Remember from our discussion of the Count-Min Sketch that it's composed of multiple arrays, each of width w and length m—pretty interesting, huh? It doesn't take many changes to go from one to the other. Similar to the Count-Min Sketch, the elements of the bit array are indexed starting at 0 ending at $(m-1)$, and because the Bloom filter is a binary bit array of length m, the space requirements are $m/8$ bytes.

Figure 5.10 shows how this algorithm works.

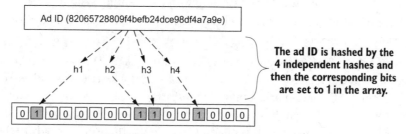

Figure 5.10   A Bloom filter showing one stream element being processed

That's it—quite straightforward. You may have already realized there will undoubtedly be collisions as you process all your data, and those can lead to the false positives I mentioned. When that happens, if the bit in the array is already set, it remains set. Querying a Bloom filter to check the membership of a stream element is also quite simple and comes down to this:

MEMBERSHIP of Stream Element Z = AND(h1(Z),h2(z),h3(z),h4(z));

With this, we compute each of the hashes and then check the array to see if all the elements are 1, and if any of them is 0 we are guaranteed that the element has never been seen before. To dig deeper into Bloom filters, a great place to start is the original article by Burton Bloom titled "Space/Time Trade-offs in Hash Coding with Allowable Errors" (*Communications of the ACM*, 1970), there are many papers that have been published since then that discuss more advanced bloom filters.[7]

---

[7] Bloom's article can be found at http://citeseerx.ist.psu.edu/viewdoc/summary?doi=10.1.1.20.2080.

This data structure can be used to determine whether you have ever seen a stream element—before incurring the cost of performing an expensive computation that may involve querying an external data store. Maybe you're building a network-monitoring application that keeps track of known bots and/or bad hosts. As you watch traffic flow you can query a Bloom filter, and if it comes back positive that the packet was from a malicious host, *then* you can perform the more costly operation to confirm if it is indeed a packet that should be rejected. Maybe you're not building either of these, but I think you get the general idea here. It may not come as a surprise that this data structure is called a *filter*, as that is the most common use case.

5.4   *Summary*

In this chapter we took a step back from discussing architecture and dove into how to think about querying a stream, considered the problems with time, and dug into four popular summarization techniques. You learned about the following:

- The different types of queries
- How to think about time when dealing with a streaming system
- Four powerful stream summarization techniques that form the basis of a lot of streaming analysis programs.

 I understand that some of this may have been a little deep. Don't worry about it. As you start to build out a streaming system, a lot of this will start to crystalize. You may want to pick one of the summarization techniques and apply it to one of the problems you're trying to solve. The architecture may be fun, but the exciting part comes when you apply what you've learned in this chapter. My hope is that you're ready to start asking questions of the data you're working with.

The next chapter covers how to store the results of the analysis you learned to perform in this chapter. This may be a good time to refill your coffee.

Storing the analyzed or collected data

Up to this point we have spent time discussing the architecture and the algorithms commonly used in stream-processing applications. This chapter focuses on what to do with the data after you have processed it. Our focus will be less on the performance of any persistent store we use and more on how to choose a persistent store if one is needed for your application.

First let's recap where we are in the overall architecture. Figure 6.1 shows our over-arching architecture with the focus of this chapter highlighted.

We'll talk about the storage options from a streaming perspective, the key attributes of the most popular products, and things to consider when you use one. To set the stage, let's begin with the four options we have when our analysis is done and the data is ready to be consumed. We can do any of the following:

- Analyze and discard the data
- Analyze and push the data back into the streaming platform
- Analyze and store the data for real-time usage
- Analyze and store the data for batch/offline access

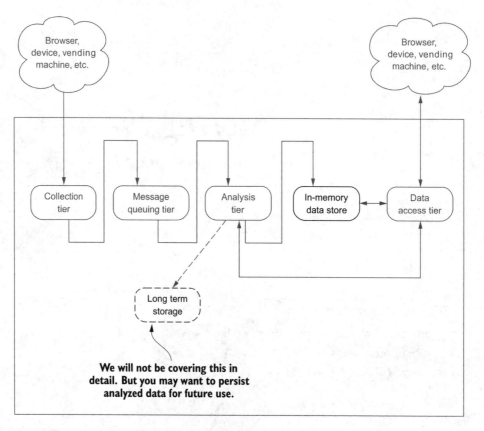

Figure 6.1   Overall streaming architecture blueprint with the chapter focus identified

Let's walk through each of these briefly. Discarding the data may be a realistic use case. You may perform an analysis of the data and subsequently discard it if it doesn't meet a certain threshold. You may not need me to remind you, but if you're only processing the data as a stream and don't have another copy of it—either from the source or from processing it via a more traditional batch process—then when you discard it, it's gone for good. This option's implementation and impact are quite obvious.

Pushing data back into the streaming platform is an interesting option. The output from one streaming analysis can become the input to another. This type of pattern is common when chaining multiple stream-processing workflows. In fact, this pattern is at the core of Apache Samza.

Then we have the last two options: analyzing and storing the data for real-time access and analyzing and storing the data for batch/offline access. Why in a book on streaming systems are we spending time discussing storing data for batch/offline access? Great question. We won't be digging into how to choose one of these data stores; many books and online resources are dedicated to that topic. Instead, we will be looking at these storage options from the viewpoint of a stream and developing our understanding of the delicate balance between the two systems. After that we will turn our focus for the rest of the chapter toward the technologies and ideas we need to keep in mind for our in-memory store. By the end you will know how to handle both storage scenarios and be poised to set up a streaming system with an in-memory store.

6.1 When you need long-term storage

There will be times when you need to have the data you're processing in your streaming system stored in a storage system designed for a non-streaming scenario, perhaps for more traditional batch or offline access. For example, writing to Amazon S3, HDFS, HBase, or many traditional RDBMSes would be considered writing to a non-streaming data store. When the need arises to write to these stores, you have the three options shown in figure 6.2.

As you'll see in later sections, some modern in-memory technologies and techniques let you write to a long-term store in different ways than described in figure 6.2.

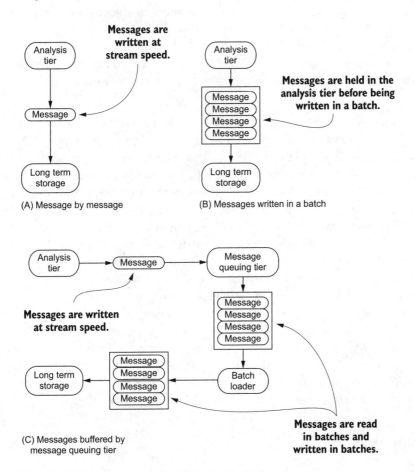

Figure 6.2   Three ways to write data to long-term storage

But for now we will focus on the options we have if we're only interested in populating a long-term store and not in-memory.

DIRECT WRITING

Start with options A and B in figure 6.2, both of which I consider to be forms of direct writing, even though they're slightly different in that the analysis tier is writing directly to the long-term store. Option A is the most direct method of getting data to a long-term store. Following this pattern, you write each message as it is processed—in essence writing at stream speed. In option B you build up the messages in the analysis tier until either a certain batch size is reached or a certain period of time passes and then write them to long-term storage.

There are potential issues and risks in options A and B. If the data store you're writing to can't keep up with the rate at which you're processing data—our stream time—then that may have a negative impact on your ability to process the data in the stream. If you're processing data at speed, it becomes harder and harder to tune a batch-oriented storage system as your data volume and velocity increases. And that makes perfect sense because these are systems that are often designed for situations where data is loaded by batch and then processed. Although option B involves building up batches of messages before writing them, it is an attempt to optimize for the slow write speed (*slow* is relative—in this case it's related to stream speed).

INDIRECT WRITING

In figure 6.2's option C the approach is quite different and falls into a category I'll call *indirect* writing. In this case we decouple the storage of the stream-processed data from the long-term store. We write the data out to a message queuing tier as an intermediary. This helps decouple our streaming analysis tier from the performance of the long-term storage system. We've added a little complexity, but that's outweighed by the benefit of offloading the data storage to a more appropriate place. The added complexity we need to take on in this case is in the form of another component that needs to be used, called in figure 6.2 the *batch loader*. The goals of the batch loader are to read batches of messages from the message queuing tier and write them to our long-term storage.

This approach has two benefits. First, as we know from earlier chapters, the message queuing tier is designed to handle the speed and volume of a stream, eliminating the risk of not being able to write fast enough. Secondly, we can use a component that is dedicated to bulk loading data, ensuring that we can keep our streaming analysis focused on analyzing the stream. Although we won't focus on bulk loading tools in great detail, it is important to understand the overall capabilities these tools provide. From a capability standpoint, the most common features that you would expect from data ingestion extract, transform, and load (ETL) tools are as follows:

- Job/task scheduling
- Error handling
- Data quality checking
- Data publishing

- Monitoring/metrics
- Horizontal scaling
- Fault tolerance
- Extensibility
- Strong consistency

Although it's easy to look at that feature set and reason about how we would have some of that in our stream processing, consider that we often want to write to a long-term store like HDFS or S3, neither of which is ideal for continuous high-volume writing, having been designed for writing large amounts of data in a batch-oriented fashion. From an architectural standpoint the long-term store is also cleaner because we have a separation of concerns and components focused on a single task.

Two of the most prominent tools used for performing this type of batch loading are Goblin and Secor. Linked-In's Goblin project (http://gobblin.readthedocs.io/en/latest/) is a universal data ingestion framework, the latest built by LinkedIn after years of experience building and running ingestion frameworks. Another popular one is Pinterest's Secor project (https://github.com/pinterest/secor). Both projects provide good options for ingesting data from Kafka or another data source and publishing it to HDFS and/or S3. If your project requires batch loading of data into a long-term store, I encourage you to consider one of these. Make sure you look closely at their feature sets and ensure that they match your requirements.

At this point we've discussed the two main options for getting data from your streaming analysis to a long-term data store. Now let's look at the in-memory options and how we can make our data available for real-time use. Get a refill and then come back and get ready to explore the world of in-memory data storage.

6.2 Keeping it in-memory

When building a streaming analysis system, the goal is to be able to take action on the data in real time, when an event occurs, as soon as possible. Imagine we're running a power company that has smart meters deployed across the world. Each meter reports its status every 30 seconds, and there are many meters connected to a single transformer. We can easily analyze the data about the meters, but what if we could take the data from all the meters connected to a transformer and watch for trends? Perhaps using a 30-minute window we see that the transformer is starting to trend toward malfunctioning and we can immediately shut it down before it malfunctions and possibly destroys other components. To be able to make this type of decision we need the current and recent data to be accessible in real time—we need it in-memory.

Not too long ago, keeping a lot of data in-memory seemed like a good idea for caching, but not for analytics. Times have changed, and the writing has been on the wall for a while. In 2006 the late Jim Gray gave a talk titled "Tape Is Dead, Disk Is Tape, Flash Is Disk, RAM Locality Is King" (http://research.microsoft.com/en-us/um/people/gray/talks/Flash_is_Good.ppt). More recently, Gartner analysts have been quoted as saying, "RAM is the new disk, disk is the new tape." There *is* still tape, but its disappearance is only a matter of time. Today you can buy servers with 32 TB or even 64 TB of memory. For approximately $5,000 a month you can have a cluster in Amazon EC2 with 1 TB of DRAM. No matter how you look at it, the idea of keeping the entire working set for your streaming system in-memory is now a reality and is something you should seriously consider as you begin to build a streaming system.

You may be thinking, "I have SSD drives—isn't that fast enough?" Perhaps, but a single seek on an SSD drive today takes approximately 100,000 nanoseconds (ns); it only takes 100 ns to reference main memory, and that goes down to 0.5 ns to access an L1 cache reference. No matter how you slice it, accessing data in DRAM is still significantly faster than even SSDs. It may seem like 100,000 ns is fast enough, and for a single access it may be. But when you're processing a continuous stream of data, all such little costs add up to be quite significant. It's no surprise that since we can now economically store entire data sets in-memory, all of a sudden every vendor has an in-memory offering, and each is better than its competitors. The question is: How do we match them to our use cases?

Let's look at the categories these technologies fall into and at some products you will find in each category. After that we'll walk through examples to help you choose among them.

6.2.1   *Embedded in-memory/flash-optimized*

See
chapter 9
section
9.4

In this category are products designed to be embedded into your software. As such they are focused toward a single node, don't provide any way to access data across nodes, and don't provide any of the management-related features you find in non-embeddable systems. These products are not a good fit for the distributed streaming systems we have been talking about building. Remember, we want to store the analyzed data so that we can make it accessible to clients in real time. Using an embedded database means we have to find a way to get access to our analysis nodes. Figure 6.3 illustrates this.

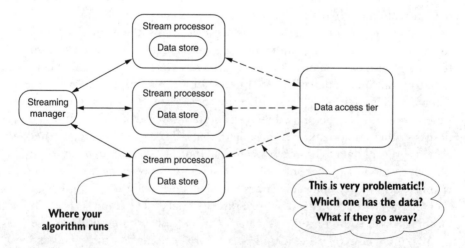

Figure 6.3   Architectural view of using an embedded data store with our analysis nodes

In figure 6.3 you can see that the embedded data store will live on the same node as our stream processor. In this case we will run into several problems that will need to be addressed, two of which are called out in the figure. It doesn't take long to realize this type of design is not ideal; it's fragile, prone to error, and doesn't follow good software architecture design patterns. The stream processor nodes are performing many roles: stream processing, local data storage, and serving data to a data access tier. For these reasons I won't spend much time discussing this approach, but I realize there may be a time and use case in which your streaming system requires this type of architecture. The following list suggests products you may want to consider:

- *SQLite (www.sqlite.org)*—This is a serverless embeddable database designed for local storage for applications and/or devices. Supports most of the common features of the SQL language; for a list of what isn't supported, see www.sqlite.org/omitted.html.
- *RocksDB (www.rocksdb.org)*—An embeddable key-value store designed for fast storage that builds on LevelDB. It can be used as the basis for building a traditional client-server solution.
- *LevelDB (https://github.com/google/leveldb)*—The precursor to RocksDB, an embedded key-value store that provides ordered mapping from keys to values.
- *LMDB (http://symas.com/mdb/)*—An embeddable key-value store developed as a replacement for Berkley DB. It is fully transactional, uses memory-mapped files, and is designed for read-heavy workloads.
- *Perset (www.mcobject.com/perst)*—A fully transactional object store for Java and .NET, designed for speed, easy of use, and transparent storage between supported languages and the data store.

This list is by no means exhaustive. You will find different options as you research this topic.

6.2.2 *Caching system*

Products in this category may also be called by other names, such as an object caching system, in-memory store, or even in-memory key-value store. The key features are that they are designed to store data in memory, there often is no option to store the data outside of DRAM, and often the API is key-value based. Caching systems use many different strategies; the ones related to persisting cache entries and keeping the cache fresh are most relevant to our discussion in this section. Let's go over the common strategies employed by most of the products in this category.

READ-THROUGH

With this strategy the caching system reads data from a persistent store when it's asked for a cached entry that isn't in the cache. Read-through incurs the cost of reading from the persistent store the first time a cache entry is asked for that doesn't already exist in the cache. This is transparent to the client of the cache, but there is the performance impact of the caching system having to read the data from a persistent store, as well as the updates having to be written to the persistent store. Figure 6.4 illustrates this strategy.

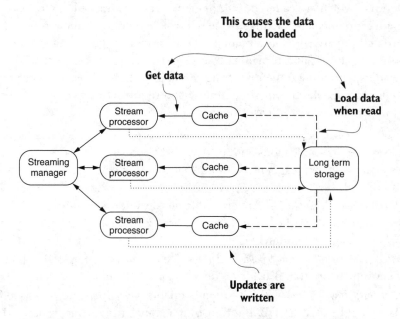

Figure 6.4   Read-through caching strategy showing how the data is loaded from long-term storage if it's not in the cache

REFRESH-AHEAD

The goal of this strategy is for the cache to refresh recently accessed data before it's expired and evicted. This attempts to avoid the read-through performance penalty of having to retrieve data from a persistent store when a cache entry expires and is evicted from the cache. If refresh-ahead is configured to closely match the update frequency of data in the backing store, then it may enable the cache to return the most current value to a client and thereby keep it in sync with the backing store. Keeping this level of fine-grained coordination may be hard to do with a stream of data. Figure 6.5 graphically depicts this strategy.

WRITE-THROUGH

This strategy has the caching system write updated data through to the backing store, eliminating the need for an out-of-band process to write data to the backing store or load changes into the cache. With this feature a caching system doesn't acknowledge the write as being successful until it's written to the backing store. With this you incur the latency of writing to the backing store, which can be deemed a disadvantage compared to the other strategies. This strategy is depicted in figure 6.6.

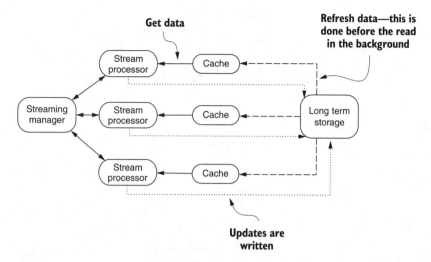

Figure 6.5 Refresh-ahead cache strategy, showing how data is refreshed from long-term memory

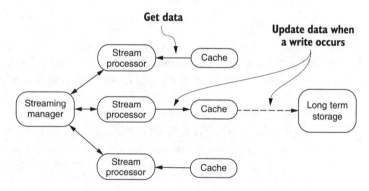

Figure 6.6 Write-through cache strategy showing how data is written all the way through to the long-term storage

WRITE-AROUND

With this strategy the idea is that the process of updating a persistent store that the cache is representing happens out of band of the cache. *Out of band* refers to the updating of the cache happening in the background—the dotted line in figure 6.6. Often out of band refers to an activity happening in a secondary pathway other then the primary. In this case the caching system doesn't know about the changes being written to the persistent store and relies on another process to update the cache after the persistent store is updated. Here you're adding the complexity of having to update two data stores: the persistent store and the cache. But the advantage is that the caching system doesn't have to talk to the persistent. Figure 6.7 shows this graphically.

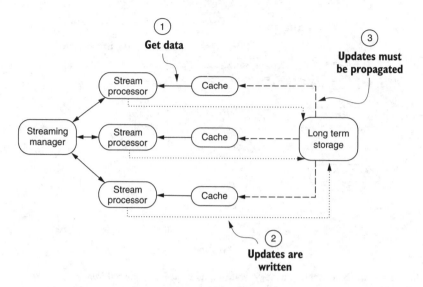

Figure 6.7 Write-around caching strategy showing how data is written to long-term storage and then has to be propagated to the cache

WRITE-BACK (WRITE-BEHIND)

With this strategy, the caching system eventually writes the new data to the persistent store. Unlike write-through, where the data is immediately written, with write-back the write to the cache is acknowledged, and then in the background the updated or new data is written to the persistent store. This has the advantage of not incurring the I/O overhead of the write-through, although for the period of time when the data is only held in memory there is a risk of data loss. Write-back is illustrated in figure 6.8.

With all these strategies you're either in a situation where the data is only in-memory or you pay the cost of having to load data from a persistent store. The fact that there is

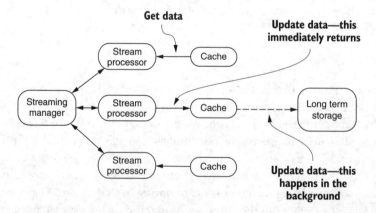

Figure 6.8 Write-back caching strategy

often no option to store the data other than in memory poses a risk as well because if you lose a cache node, you lose your data.

Some systems in this space, such as EHCache (www.ehcache.org), do provide ways of extending it to support read-through, write-through, and write-behind functionality. With a streaming system processing a constant flow of data, this dual writing and propagation of changes will become a significant bottleneck. If your results are transient and never need to be persisted, this bottleneck may not be an issue. In general, though, you'll want to look at historical data—in which case, using a caching system will force you to dual write your streaming analysis results, potentially incurring significant overhead.

If you're interested in exploring these systems in greater detail, here are several of the most popular open source products in this segment:

- *Memcached (http://memcached.org)*—This is a popular cache product, but you'd need to employ the write-around strategy to keep it fresh with data.
- *EHCache (www.ehcache.org)*—This is a sophisticated caching system with support for various usage scenarios—namely write-behind, write-through, and read-through.
- *Hazelcast (http://hazelcast.org)*—This is a robust product with a lot more than caching. On the caching side, it has support for read-through and write-through strategies.
- *Redis (http://redis.io)*—This is a popular option that works well as an in-memory cache. It does have persistence but to its own file formats. You'd need to build any of the strategies discussed earlier.

6.2.3 *In-memory database and in-memory data grid*

In-memory databases (IMDBs) are sometimes called *in-memory database management systems* and *in-memory data grids (IMDGs)*. Unlike the caching systems, IMDBs and IMDGs are designed to use disk for the non-volatile persistence of data. Although the products in this category use a storage medium other than DRAM, they are designed to use memory first and disk second. In fact, disk storage is often only used for logging and periodic snapshots so the data is available in the face of failure. This subtle design decision is what separates the true IMDBs and IMDGs from the traditional databases that now offer an in-memory option. Traditional databases (Microsoft SQL Server, Oracle, and so on) as well as modern NoSQL ones (such as Apache Cassandra) were all designed with a disk-first, memory-second mindset.

What is a disk-first mindset versus a memory-first mindset? Glad you asked. Figure 6.9 illustrates the stark differences in the approaches.

The disk-first approach shown in figure 6.9 is a simplified picture of how a traditional database and many NoSQL databases have been designed. It should be apparent why products that started as disk-based systems and now claim to have in-memory features were designed this way. The problem is revealed later: When that product adds features to become an in-memory database, the same design principles are applied;

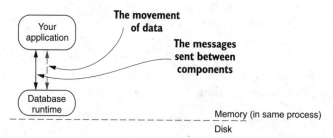

Figure 6.9   The two different approaches to database design: disk-first and memory-first

memory is swapped for disk but treated like another I/O device. That's not to say these products don't use memory from the start; for performance reasons they all happily use memory as a cache.

The memory-first mindset as shown in figure 6.9 completely flips this. With the memory-first approach, the software is designed to only use memory, and then later the disk is used secondarily for reliability. This subtle difference has a profound impact on the performance of the products in this segment. Several products cross over into this category from the caching category; with IMDB and IMDG products, the lines are getting blurry, change frequently, and are at this time fairly fluid.

Traditionally there were several key differences that distinguished between IMDBs and IMDGs. IMDGs used a distributed architecture and allowed computations to be performed on the server close to the data, similar to the stored procedures you're used to in the RDBMS world. IMDBs would often provide API access via SQL, whereas IMDGs usually used a non-SQL API. As the industry has matured in the streaming space, these lines are getting erased, and IMDGs now sometimes find themselves having to offer much richer computation models as products like Apache Spark encroach on the in-memory computation capabilities. As this has been happening, IMDBs have been moving closer to being IMDGs, and IMDGs have been moving toward trying to compete with streaming analysis tools and complex event-processing frameworks. A few of the most common open source products you'll find in this segment are as follows:

- *MemSQL (www.memsql.com)*—This started as an in-memory-only option but now also supports storing data on disk. The main idea is that it is 100% MySQL compatible, but much faster because by default all data is in-memory.
- *VoltDB (www.voltdb.com)*—This is a high-performance SQL-compliant database that offers an in-memory option as well as options for durability, high-availability, and export integrations.
- *Aerospike (www.aerospike.com)*—This is a flash-optimized NoSQL engine with a lot of different features, ranging from geospatial indexing and querying to geographic replication.
- *Apache Geode (http://geode.incubator.apache.org)*—This was originally designed to be in-memory but also offers off-heap storage and a query language called OQL (Object Query Language) that is SQL-like.
- *Couchbase (www.couchbase.com)*—This document database was originally a mix of Memcached and CouchDB but has grown well beyond that. It can be geo replicated and has its own query language called N1QL (pronounced "nickel") that is a SQL extension for JSON data.
- *Apache Ignite (https://ignite.apache.org)*—Billed as an in-memory data fabric, in the end Ignite is an IMDG with the addition of a rich compute grid and SQL support and integrates with Hadoop and Spark. Although a young Apache project, it started life as a product offering from GridGain and was subsequently donated to the Apache Foundation.
- *Hazelcast (https://hazelcast.org)*—This is an IMDG that offers features (at least at the time of this writing) such as queuing, distributed aggregations, Map/Reduce support, and distributed data structures.
- *Inifispan (http://infinispan.org)*—This is an IMDG that also offers features such as integration with Apache Spark and distributed streams and performing complex operations via script.

6.3 *Use case exercises*

We've covered a lot of ground, going over different aspects of in-memory storage. Given all the product options and the never-ending list of new contenders, it's impossible to discuss all of them. But we're now ready to walk through several use cases and tease apart which product category may suit our needs. Without a doubt your requirements may be drastically different than those for our use cases, but my hope is that you will be able to take the information and apply it to your use case. I know I haven't talked about the data-access side of streaming analysis, so the upcoming uses cases keep the focus on making the data available for being accessed and may make some assumptions about how it is accessed.

6.3.1   *In-session personalization*

For this use case, imagine that we've built a stream-processing system behind a popular e-commerce website called TheOceana.com, where the product catalog is as vast as the world's oceans. We have a stream of all activity on our site and we want to personalize the site for our users while they're in the midst of shopping. This is often termed *in-session personalization*, because it's occurring during the user's current session. Our goal is to change the page content at request time based on the activity in the user's current session. Imagine a user's session looked like this:

1   Browsed sunglasses
2   Added yellow sunglasses to cart
3   Browsed motorcycle helmets
4   Removed sunglasses from cart
5   Browsed jackets
6   Browsed motorcycle gloves

Now, we want to personalize this experience in the following ways:

- When they go to the motorcycle helmets in step 3, we want to show them helmets that would go with their yellow sunglasses.
- When they start to browse the motorcycle gloves in step 6, we want to show them a coupon for the sunglasses they removed from their cart a couple pages ago.

Figure 6.10 shows this flow along with the personalization actions we want to take.

I'm sure you can think of many other ways to personalize the page, perhaps not based on a single user's session but maybe for all active users, or maybe you can think of ways to combine the current session with the visitor's history. For now, let's focus on the two ways we want to personalize the page content: first by showing related content and then offering a coupon for a product that was removed from their cart. To do this we need to keep track of and analyze the active visitor sessions. In our scenario, a visitor session has a lifespan of 30 minutes and contains each action taken by a user on the site, often called *click-stream data*. Think about how you might solve this problem using each of the in-memory options we've discussed. All right, let's walk through each of our in-memory options and see how we can solve the problem.

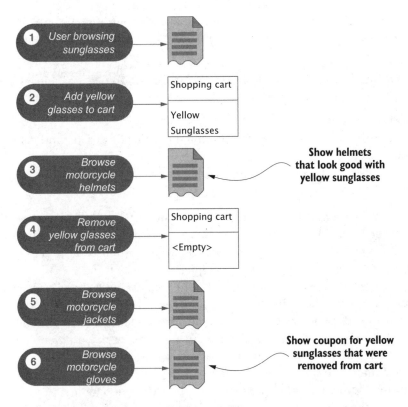

Figure 6.10   In-session personalization page flow

Storing the data in-memory on the analysis nodes presents some challenges for this use case. Namely, we need to ensure several things to be able to make this work correctly:

- We need to store the entire session for all visitors across all the analysis nodes.
- Our analysis nodes will need to know when a session ends and how to evict the data from memory.
- We need a way for a client to query our analysis nodes to get a visitor's session. Unless we had a way to perform a distributed query, the client would need to know which node has a particular visitor's session.
- We need a way to ensure that when the analysis node storing a visitor's session crashes, we won't lose all the data.

Considering these requirements and/or risks, to use an embedded database we'd need to not only satisfy the requirements, we would also need to mitigate the inherent risks of storing the data in memory. Therefore, it's not a good fit. In fact, if you spend time trying to handle them with any of the leading stream-processing products we've looked at—Apache Storm, Spark, Samza, or Flink—you'll quickly realize that the design is fragile, not scalable, and from an architectural standpoint not clean at all. I would say this type of storage option is not a good fit.

CACHING SYSTEM

Would a caching system be a good fit for this use case? To use a caching system we need to ensure at least the following:

- That we don't lose data if a cache node crashes
- Clients can query the cache cluster using the visitor ID as a key and get back the entire session
- Our analysis nodes can constantly update the session
- The session expires from the cache after 30 minutes

Can we accomplish all of these? Ensuring that we don't lose data when a cache node crashes means we need a caching system that redundantly stores data across nodes. Many products in this category don't support replication, so this may be a risk we can't mitigate. Providing the ability for a client to query the cache using the visitor ID is easy for all caching systems because they operate with a key-value API and will return the data stored at a given key—in this case, the visitor's session. With the continuous updating that needs to be done to this data, we have to consider that we won't only be replacing a value stored with a key—we need to append new data to the value. In that case, how do we handle updating a visitor session as they are browsing our site? We are left with one option because we'll need to use the visitor ID as the key. We will need to constantly read the entire visitor session from the cache, update it, and then write it back to the cache. This is a lot of thrashing that will undoubtedly have a significant impact from a performance standpoint. Perhaps we can do better.

IMDB OR IMDG

What about an IMDB or IMDG? Would it be a good fit for this use case? Consider the same basic requirements we saw for the caching system:

- We need to ensure we don't lose data if a single node crashes.
- We need to provide a way for a client to be able to query the system for the visitor's session using the visitor ID as a key.
- We need a way for our analysis nodes to constantly update the session.
- We need a way for the session to be expired from the cache after 30 minutes.

The first one, ensuring that the data doesn't disappear when a single node crashes, is easy for many of these products because they seamlessly take care of scaling horizontally and ensuring that data is replicated across nodes in the cluster. Querying the cluster using a visitor ID to get back the full session is also pretty straightforward using many of the IMDB or IMDG products on the market. The last requirement—being able to expire a session after 30 minutes—can be accomplished easily with Aerospike, which acts like a NoSQL store that has the ability to add a "time to live" value for a record that is written. With other IMDBs or IMDGs, specifically MemSQL and VoltDB, you need to determine the session expiration outside of the data store and subsequently delete the data. Overall, satisfying this use case using an IMDB is fairly straightforward, and the technology is a good match.

TAKING IT TO THE NEXT LEVEL

Some of our choices may have seemed quite easy. Taking it one step further, would your assessment change if we also wanted to support the following two additional requirements?

- Instead of using a single visitor's session, we wanted to take into consideration all visitors' sessions for that current day.
- Along with the visitor's current session we also wanted to include their entire visitor history when making our decisions.

Next we'll set up another use case, but this time I'll ask you to work through the questions independently.

6.3.2 *Next-generation energy company*

Imagine we're building a next-generation energy company, one that can help avoid brownouts and blackouts that plague certain U.S. states during the summer months. Utility companies have long suffered from this type of Saturday afternoon scenario: it's hot out, and all our customers are running their air conditioners. They also go to the fridge to grab a cold drink, and since they're in the kitchen they decide to wash the dishes. It's so hot outside they keep doing things inside. Perhaps they start a load of laundry and turn on the TV. As you can imagine, all of this spikes the demand on our energy company, and when this demand exceeds our capacity, trouble ensues.

We decide to change this: We're going to build a smart grid and let our consumers participate in the solution. Our first step is to deploy smart meters at our customers' homes. These meters will report the energy use of the home every five minutes, along with information on which appliances are consuming the energy. From all this data now streaming in, we would like to offer our customers variable pricing in real time, based on when and how they run their major appliances. If they run their air conditioner at a slightly warmer temperature or run their dishwasher at night, for example, we would offer them a discount.

To succeed, we need to perform streaming analysis in two places: at the power plant and at the home. First let's consider the features we need for the power plant:

- Ability to analyze the data, taking into consideration weather and historical data, and store it every five minutes.
- Ability to store pricing for the next hour of use (perhaps we decided to offer hourly pricing rates).
- A way to query the data store for the pricing information for a customer.
- A way to ensure we cannot lose customer data.
- A way to ensure this data is available for further analysis.

Take a moment and work through these questions for each of the in-memory data stores we discussed. How would your decisions change if you were to move the processing of this data to the meter? It may seem far-fetched, but a system like this is where energy is headed.[1]

6.4 *Summary*

In this chapter we looked at different options for storing data in-memory during and after analysis. We didn't delve into the use of disk-based, long-term storage solutions because they are often used out of band of a streaming analysis and don't offer the performance of the in-memory stores.

In this chapter we:

- Learned about the different types of caching approaches
- Compared different types of in-memory data stores
- Discussed how to choose the right one

Some of this may have seemed like we're only seeing half of the picture because part of choosing the correct store also involves taking into consideration how we access data. Chapter 7 focuses on that aspect of accessing and making available the data that we have stored. The key takeaway from this chapter is that many different in-memory options are available, and the idea of keeping your entire working set in-memory is no longer a dream; with modern hardware, software, and the continuous drop in pricing, it is a reality today. When you're ready, go ahead and turn the page and let's get going on accessing the data we've stored.

---

[1] A good starting point to read more about this is a publication from the U.S. Department of Energy titled *The SmartGrid: An Introduction*, www.smartgrid.gov/files/The_Smart_Grid_Introduction_200804.pdf.

Making the
data available

7

This chapter covers:

- Common communication patterns
- When to use webhooks, HTTP Long Polling, server-sent events, and WebSockets
- Use case: building a Meetup RSVP streaming API

We have come a long way through our architecture and are now ready to consider how to deliver the data to a streaming consumer. When designing this tier, we are faced with a similar problem as with the other tiers—that there are myriad technologies we can choose from and many ways we can build it. The other tiers have dealt with ingesting data, moving it around, analyzing it, and getting it ready for use. Without a doubt they all have their challenges and are fun and exciting to build, but I find the data access tier the most rewarding and fun of all. It is where all our hard work pays off—by delivering data to a client application. The crux of this tier is that we enable our customers to build applications with real-time features using our API. Many benefits come from having access to this type of API. One that quickly comes to mind is customers may derive quick business value by being able to see and act on data in real time. Another is by enabling developers to build more compelling applications, you can improve the development experience and empower them.

Our goal in this chapter is to see how we can build a streaming data API, taking into account the various communication patterns, discuss how to handle failure when we are building it, and look at the numerous protocols available. Figure 7.1 shows where we are in relation to the overall architecture we're working our way through. We'll begin our discussion with communications patterns.

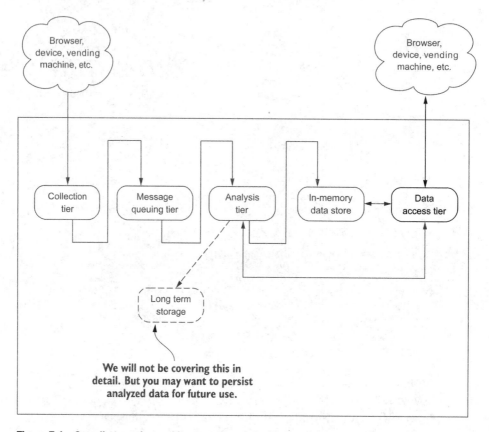

Figure 7.1   Overall streaming architecture blueprint with the chapter focus identified

7.1   *Communications patterns*

With a streaming system we're handling a continuous stream of data that is often being pushed at us, but when it comes to the communication patterns with clients, this often doesn't hold true. We need to think about supporting pushing data to a client and/or the client pulling data from our API. Beyond generic push or pull, there are four common patterns: Data Sync, RMI/RPC, Simple Messaging, and Publish-Subscribe. These are loosely based on the patterns you may see in the message queuing tier as well. Let's briefly walk through each.

7.1.1　*Data Sync*

It may seem like any communication between our API and the client involves syncing data, but with the Data Sync communication pattern, a database or data store is often synchronized between the API and the client. Figure 7.2 shows the flow of how this commonly works.

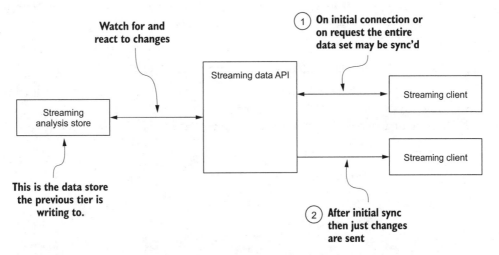

Figure 7.2　Data Sync communication pattern showing initial and subsequent synchronization

The general flow of this pattern is that the streaming API watches (there may be a notification mechanism) for changes to the data store the analysis tier is writing to and then sends the updates to the streaming client. As shown in figure 7.2, there are usually two steps to this pattern. First, when the client initially connects—this may be when a mobile application is installed, for instance—the client will request the current data set. Subsequently, changes to the data set are either pushed to the client, or the client pulls the data. This pattern may seem simple enough, but there are benefits and drawbacks that you should consider before diving in.

BENEFITS

- Simple protocol.
- The client has a complete data set.
- The API is straightforward to develop because you are only supporting an initial sync and a delta after that based on either time or perhaps a version of the local data.
- The client always has a consistent view of the data that is the most current.

DRAWBACKS

- The data set may be large and require significant bandwidth to transfer.
- The data set may not fit on the device it is being transferred to.

- Need to resolve data version conflicts.
- Need to determine a merge policy so you know how to handle data changes made by the client with those made by the server.

Even with the drawbacks, for many applications this communication pattern may be beneficial and a great fit. For example, for a multiplayer mobile game application—where you want to always have the state of the game represented on every device without worrying about treating every move as an individual transaction—moving the current dataset may make sense.

7.1.2   *Remote Method Invocation and Remote Procedure Call*

With the Remote Method Invocation (RMI)/Remote Procedure Call (RPC) communication pattern, the API server invokes or calls a method on a connected client when new data arrives or when a condition is met that is of interest to the client. The general flow of this is depicted in figure 7.3.

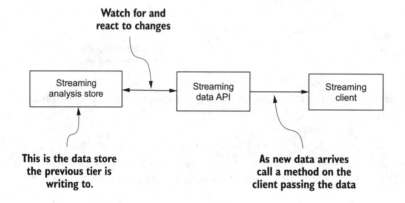

Figure 7.3   Overall flow of RMI/RPC communication pattern

As you can see, the RMI/RPC communication pattern is fairly straightforward. In general, our API monitors the data store being written to by the analysis tier for changes and then sends them to the client via remote method calls. In this case, either data is being sent as part of the call, perhaps the latest value, or it may be a call to notify the client that a certain condition has been met.

BENEFITS

- Simple protocol.
- Client can perform other processing and then react when a handler is called.
- API is straightforward to develop—all you need is a way for a client to register an endpoint.

DRAWBACKS

- Detecting failures is hard—what if the client isn't available? How can the client know the server didn't receive new data?
- Frequency of updates may overwhelm a client.
- How does the API handle client failures?

7.1.3 Simple Messaging

With the Simple Messaging communication pattern the client initiates a request to the streaming API asking for the most recent data, and the API responds with the latest data. Without adding metadata to the request to indicate to the server "only return data if it is newer than X" or the streaming API letting the client know that the data won't change for X period of time, this pattern is inefficient because the client may be requesting data that hasn't changed since it last made the request. Continuously requesting the same data that hasn't changed is a waste of resources and an inefficient pattern. Later in this chapter, when we talk about the various protocols, you'll see how this simple pattern takes a form that makes it more efficient. Each form adds a little something to the request and/or the response so that the client doesn't make extra calls to the streaming API for data that hasn't changed. Figure 7.4 shows the general flow of how this pattern works.

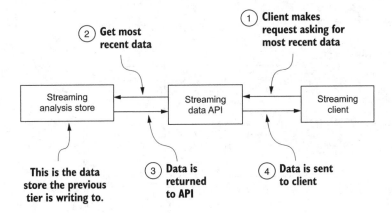

Figure 7.4 The Simple Message communication pattern with client asking for most recent data

As you can see in figure 7.4, the streaming client asks for the most recent data, the API queries the streaming analysis store, the data is returned to the Streaming Data API, then to the client. There can be subtle variations to this. For instance, the client may be able to make a request asking for changes from a certain time forward. In that case the API can check for changes and return the data changes since that point—or return no data at all if there were no changes since the time specified by the client.

- Simple protocol and API call for the consumer to make.
- Only the most recent data is sent to the client.
- The client only has to keep track of a little metadata to continue getting the most recent data.
- The API doesn't have to keep track of any client state.

DRAWBACKS

- The protocol can be chatty because the client may continue making constant calls for the new data.
- There is no mechanism to alert the client to the existence of new data.
- The payload of new data may be large if the client was offline for an extended period of time or the velocity of the data is large.

7.1.4   *Publish-Subscribe*

With the Publish-Subscribe pattern the client subscribes to a particular channel, and the API then sends messages to all clients subscribed to that channel when the data changes. I'm using the term *channel* here to loosely represent the idea that the data can be grouped into categories or topics. The general flow of this pattern is shown in figure 7.5.

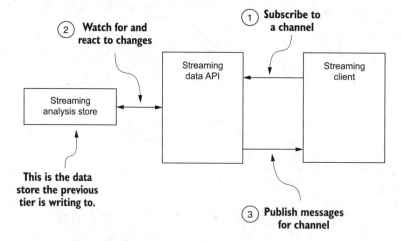

Figure 7.5   The Publish-Subscribe communication pattern

If you're familiar with the Observer design pattern, you may recognize the data flow in figure 7.5. There is an object that keeps track of a list of dependents and state. When the state changes, the object notifies the list of dependents. With the Publish-Subscribe pattern, the streaming API maintains a list of streaming clients, and when the data changes in the streaming analysis store it notifies all the streaming clients.

This pattern has many advantages over some of the previous patterns. Perhaps the biggest advantage is that the streaming API can keep track of all the clients that are subscribed to a particular channel and then publish messages to all of them when there is new data. This reduces the burden on both the clients and the streaming API. The clients don't have to constantly call asking for new data or keep track of any metadata that would inform them of when new data may be available. The streaming API is also relieved of having to respond to client requests for data that hasn't changed. That may seem like it shouldn't be a burden for the streaming API, but think about the impact of millions of clients making requests for data that hasn't changed—that would be a waste of network resources and taxing to the streaming API to respond to.

This pattern is becoming more and more pervasive as we move toward a world of more reactive programming and streaming systems. If you use Twitter and follow a hashtag using a Twitter client, you've seen this pattern in action. Slack, the instant messaging application, also uses this pattern. If you haven't used Slack, the basic feature is this: Slack allows users to join a channel and then sends updates to each subscribed user as new messages are posted. You can probably think of many other examples of this pattern in use today.

BENEFITS

- The client can perform other processing and then react when data arrives.
- The client doesn't need to maintain any metadata about the current data.
- The API can optimize how it handles sending data to multiple clients.

DRAWBACKS

- There's a more complex protocol for the API.
- The API has to keep track of all metadata related to the clients, and must have the ability to distribute this across API servers in the event of a failure.

7.2 Protocols to use to send data to the client

Understanding the common patterns of communication is a great start. But before we can build our streaming API, we should look at the protocols that are commonly used and see which ones we may want to consider for our streaming API. We'll consider the following factors for each protocol:

- Message update frequency
- Direction of communication
- Message latency
- Efficiency
- Fault tolerance/reliability

7.2.1 Webhooks

Webhooks have been around since 2007. Though not an official W3C standard, they've been adopted by many as a way for a client to register a user-defined HTTP endpoint to be called when new data arrives or a condition is met. Conceptually, this is similar to the *callbacks* you may be familiar with in many programming languages. There are some stark differences, though. First, the callbacks in this case are executed via an HTTP POST. Secondly, often the client is implemented by a third-party developer. The general way in which this works is illustrated in figure 7.6.

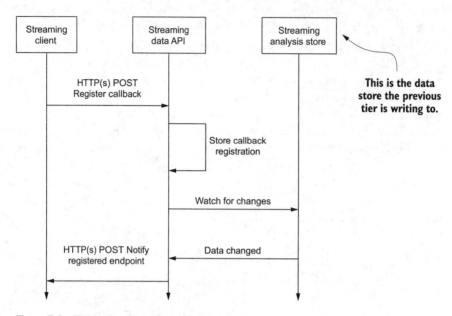

Figure 7.6   Webhook registration and callback

In the first step the client sends a request to register the callback via HTTP POST. Sometimes this step is done manually by a user filling the information in on a website. Subsequently the callback information is stored. After that, the streaming API calls all the registered callbacks when new data arrives or an event condition is met.

Now let's look at common factors:

- *Message update frequency*—Considering that every update is sent via an HTTP POST, it's fair to say that the frequency of sending updates would be low. There is nothing inherent in the protocol that would prevent a streaming API from calling it at a high rate, but considering the textual nature of the protocol and the inherent overhead, it's not ideal for high frequency.
- *Direction of communication*—The direction of communication is always from server to client. If any changes need to be made, it involves the client changing their registration. There is nothing standard about the registration process; it is 100% up to the provider of the streaming API.
- *Message latency*—The latency of messages is average. Many software stacks have highly tuned HTTP stacks, and although the protocol is text-based, we can take advantage of HTTP compression and chunking if the data updates are large.

- *Efficiency*—From the streaming API perspective this protocol is efficient. No state has to be maintained outside of the list of callback endpoints. When an update is to be sent (data or a condition), the server can make an asynchronous HTTP POST request to each of the registered callback endpoints.
- *Fault tolerance/reliability*—The protocol doesn't provide any guarantees—it's all up to us. We need to consider what to do if the HTTP POST fails. And because the communication is unidirectional, there's no way for the client to acknowledge that data was received. The HTTP protocol would allow us to make a separate GET request to see if a POST was successful, but that adds complexity because we now also need to handle failure of that request. If we choose to use this protocol to implement our streaming API, we have to address at least the following:
  - What do we do with the messages that were going to be sent to a client and the POST failed? Do we retry sending the message?
  - Is there a way for a client to get the messages that it would have missed if the HTTP POST failed?

Webhooks is a fairly simple protocol. Based on the discussion of the factors we're considering, it's not something that supports everything we would want in a streaming API. A common use case for webhooks is in systems with a low volume of messages that aren't impacted if a message is missed.

7.2.2 HTTP Long Polling

HTTP Long Polling involves the client making a connection to the server (in this case the streaming API server), the connection being held open, and the data being sent to the client as it is available. The general flow of control is shown in figure 7.7.

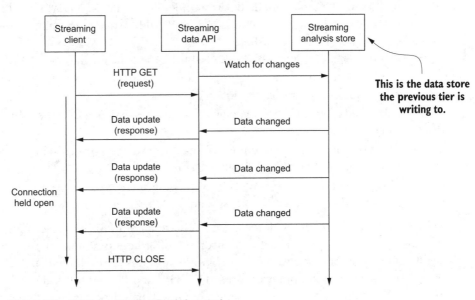

Figure 7.7   HTTP Long Polling flow of control

The client makes a request, and the server holds the connection open till there is an update and then sends it. The client in turn immediately opens a new connection, and the cycle repeats. This allows the client to control querying for data changes, but it comes at the cost of our streaming API being forced to keep open connections to all the clients.

Let's consider our common factors to see what else we need to keep in mind if we want to build our streaming API with this protocol:

- *Message update frequency*—Similar to the other HTTP-based protocols, there is nothing inherent in the communications protocol that would prevent using HTTP Long Polling for high update rates. Nevertheless, considering the textual nature of the protocol and the inherent overhead of a client getting a message, the connection being closed, and the client having to re-establish the connection, the overhead would become expensive if there was a high update frequency.

- *Direction of communication*—The client establishes the connection and can specify what it wants updates for, and the server responds.

- *Message latency*—As with webhooks, the latency of messages going over HTTP is average.

- *Efficiency*—Although the underlying HTTP protocol we are sending data over may be efficient, HTTP Long Polling isn't an efficient protocol. Our streaming API server will have clients keeping connections open to us while they wait for data and then immediately reopening them again as soon as they get a message. Our API server will be constrained by the number of open connections. We'll also need a way to ensure that if a client disconnects, the connection is closed. For example, when a mobile device rapidly switches between Wi-Fi and cellular networks or loses its connection, and its IP address changes, will a connection be automatically re-established?

- *Fault tolerance/reliability*—Similar to webhooks, this protocol doesn't provide any guarantees; it's all up to us. To implement our streaming API with this protocol we have to address at least the following:
  - We need to ensure that any new messages that arrive before the client reconnects are not lost.
  - To prevent a single streaming API host from getting overwhelmed, we need to understand how to load balance connections.
  - To prevent a client from losing or missing messages, we need to ensure we can handle the failure of our streaming API servers.

HTTP Long Polling is much closer to what we are looking for in a streaming API. It became popular with the rise of asynchronous programming in client-side development, rising in prominence with the use of Asynchronous JavaScript and XML (AJAX). The first web-based chat applications used HTTP Long Polling for real-time communication. It's still in use and is accessible to non-web-based clients as well because it's an HTTP request. It's supported by most if not all programming languages and is available on devices from servers to the small Raspberry Pi.

7.2.3 *Server-sent events*

Server-sent events (SSE) was developed and the subsequent W3C recommendation established in 2015 as an improvement over HTTP Long Polling. SSE helps solve at least two problems. First, it solves the inefficiency that exists when the client constantly closes and opens connections for every message received. Secondly, when using a resource-constrained device such as a mobile device it supports using a push proxy server, allowing the device to enter sleep mode while idle and be pushed messages from the proxy. This results in significant power savings for the device compared to keeping the connection open while idle. The two scenarios for SSEs are illustrated in figures 7.8 and 7.9.

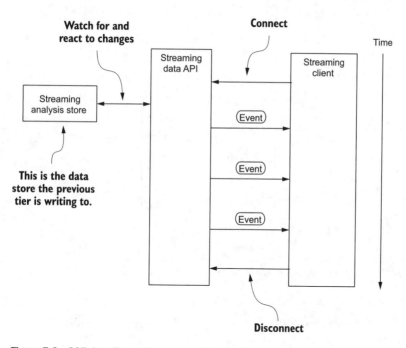

Figure 7.8 SSE data flow with a connected client

Figure 7.8 shows that the streaming client first establishes a connection to our streaming API server, and then as messages become available they are sent to the client. Unlike Long Polling, where the connection is closed and reopened for every message, in this case all the events are sent over the same connection. This results in more efficient network utilization and allows the client to do other processing while waiting for events to arrive. This still requires a client to maintain the connection, though. The second mode that SSE supports is *connectionless push,* illustrated in figure 7.9.

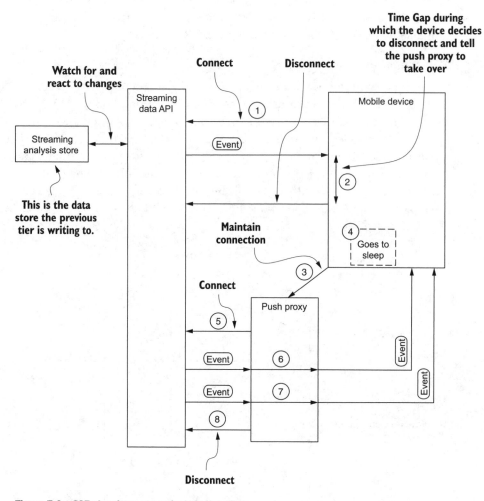

Figure 7.9   SSE showing connectionless data flow

The data flow shown in figure 7.9 is a little more complex than figure 7.8. Let's walk through it:

1   The client, in this case a browser on a mobile device, connects to the streaming API.

2   After receiving a single event, enough time goes by that the device decides (this is up to the implementer) that to save power it's going into sleep mode.

3   Before going into sleep mode the mobile device sends a message to the push proxy asking it to maintain the connection. It may send along the ID of the last event it received so the push proxy can pick up from there.

4   The mobile device sleeps to conserve power.

5   The push proxy establishes a connection to the streaming API.

6   When the push proxy receives an event, it sends it to the mobile device using a handset-specific push technology. The device wakes to process the message and then resumes sleeping.

7   This is the same flow as step 6.

8   At some point the push proxy disconnects from the streaming API, and the whole flow is finished.

For added complexity, using the push proxy, the mobile device can save power and reduce data usage using this workflow. The data use reduction and power savings are by-products of the fact that the push proxy is maintaining the connection with the API server, not the mobile device. The device doesn't pay the heavy cost of keeping a TCP connection alive, which can be expensive. When there is new data, the proxy uses a technology such as a push message to wake the device so it can receive and process the message and then go back to sleep. The result is that your users have a better experience when using your streaming API.

Considering how similar these two variants are, I'll treat them as one when looking at our common factors next, calling out any differences:

- *Message update frequency*—Although the messages with this protocol use HTTP as the transport protocol, the update frequency can be much higher than HTTP Long Polling. This increase in message update frequency is because that with this protocol there is a single persistent connection. You need to be cautious, though—you may run into problems with a client not being able to read data fast enough from the network at high message rates.

- *Direction of communication*—Outside of the initial connection, this protocol is clearly unidirectional, with the server pushing the events to the client.

- *Message latency*—As mentioned in previous HTTP protocol–based sections, the latency of messages going over HTTP is average.

- *Efficiency*—This protocol is as efficient as it can be given that it uses HTTP as the transport. There's a single connection per client over which all data is sent, eliminating the cost of constantly opening and closing connections. On the server side any gains in efficiency are in how fast data can be written to the socket, and care needs to be taken to ensure the client can read the data at the same rate off the wire. There is nothing inherent in this protocol that allows for the negotiation and managing of a maintainable message rate. Therefore, if you find that your clients can't keep up with the rate of messages, you may want to use a different protocol.

- *Fault tolerance/reliability*—Unfortunately, this protocol is similar to the previous HTTP–based protocols, where there are no guarantees. Because this protocol is unidirectional, from the server pushing to the client, we can't make this 100% fault tolerant and reliable from a message standpoint. But we can still address the following:

  - We can ensure that as new messages arrive we don't write them to the client if the network buffers are full. This would indicate the client can't read the messages fast enough. With this protocol there's no way for the client to communicate that to the API; we have to rely on gathering this information from the network stack.

– Because clients have persistent connections to the streaming API, it's important to ensure a single streaming API host doesn't get overwhelmed. Therefore, you need to know how to load balance connections.

– To ensure that our streaming API doesn't miss sending messages to a client, we also need to ensure that when a streaming API server fails, another server can pick up sending messages from the last message sent forward. There are two sides to being able to do this: first, the streaming API must keep track of, in a distributed fashion, the last message ID sent to each client. When a server fails, the server that picks up the work can continue where the failed node stopped. Secondly, the client can help by sending the last event ID it received when it re-establishes a connection. An API server can then use this information to send messages from that point forward.

Overall this protocol yields better performance and allows us to provide a more resilient API. It is becoming the go-to alternative to HTTP Long Polling and is one you may want to consider as long as the fault tolerance and reliability concerns are acceptable to your use case.

Next we'll talk about WebSockets, which provides more flexibility than what we've seen so far.

7.2.4 *WebSockets*

WebSockets has been around since 2011. It's a full-duplex protocol that uses TCP for the communication transport. All major desktop and mobile browsers support it. Although it's commonly used between web clients and web servers, libraries are available for all major programming languages, allowing more interesting use cases to be solved. WebSockets is an interesting protocol in the sense that it uses HTTP for the initial handshake and protocol upgrade request and then switches to TCP. Figure 7.10 illustrates the WebSockets data flow.

Several things are going on with this protocol that are called out in figure 7.10:

1 This initial sequence of events—the handshake—happens between the client and the server over HTTP. It starts with the client initiating the handshake as an HTTP upgrade request. The server responds to the client, completing the handshake, and the protocol is upgraded from HTTP to TCP.

2 As events arrive, the server sends them to the client over TCP.

3 The "slow down" here is because the communication is bi-directional and the client may send a message to the server at any time. If a message is sent, what it contains and means is completely up to you. In the earlier example, we can imagine that the client can't read the data fast enough and is asking the server to slow down the frequency at which messages are sent.

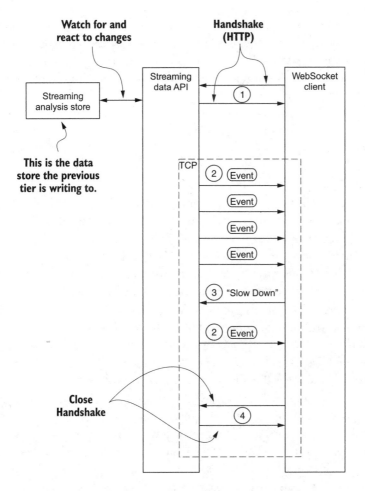

Figure 7.10 Example WebSocket data flow showing the client sending a message

4 At some point the client or server will want to end the connection. Although a connection may be abruptly closed, the correct way to take care of it is via *close handshake.* Figure 7.10 shows the client initiating the close, but the server can initiate it as well. Regardless of which side initiates it, the flow is the same: there is a close handshake that starts the close, followed by a close handshake response. At that point, the underlying TCP connection is closed.

See code listing 9.3.1

A lot is going on with this protocol, as you can tell. Let's look at the common factors to see how they can be addressed with this protocol:

- *Message update frequency*—Considering that all data for this protocol is sent over TCP, the message update frequency can be much higher than all the previous protocols we've looked at. Similar to SSEs, this increase is due to the persistent

connection and in this case the protocol. There is the same risk with this protocol as there is with SSEs, that you need to be sure a client can keep up with the
update frequency. The throttling of messages and slowing down isn't part of
this protocol, but as shown in figure 7.10 this protocol lets you implement mes-
sage-throttling semantics.

- *Direction of communication*—Unlike all previous protocols, this one is truly
  bi-directional from beginning to end, enabling you to create interesting sub-
  protocols.
- *Message latency*—All communication for this protocol is over TCP, so latency is
  low compared to the previous HTTP-based protocols.
- *Efficiency*—Given that this protocol sends all data over TCP and there is a single
  persistent connection between a client and a server, this protocol is very efficient.
- *Fault tolerance/reliability*—Unfortunately, like the other protocols we've considered, this one doesn't offer any guarantees. But unlike the others this one is bi-
  directional, allowing us to build fault-tolerant and reliability semantics into it.
  In doing so, we need to consider how to handle the following:
  - If we are receiving events and trying to write them to a client faster than it can
    read them off the wire, we need to buffer them and not write them. Leverag-
    ing this protocol's bi-directional nature, we can make the client an active par-
    ticipant and have it send a message, perhaps something similar to "slow down,"
    indicating that it needs the server to send messages at a slower rate.
  - Similar to SSEs, to ensure that our streaming API doesn't lose messages, we
    need to keep track of which message was sent to which client. Therefore,
    when a server fails and/or a client reconnects, the next message is the one
    expected and there's no gap in the messages sequence. Nothing inherent in
    the protocol provides this, but it's relatively easy to put in place. An advan-
    tage this protocol has over SSEs in this area is that with a little work you can
    have the clients send an acknowledgment to the server that not only did they
    receive the message, they have also processed it. This allows you to build a
    system that is highly fault tolerant and reliable.

After considering the common factors and the previous protocols we looked at, it may
not come as a surprise that this protocol is more efficient, powerful, and flexible than
the others. Here are some reasons for this:

- A single TCP connection is held open for as long as the client wants to consume data.
- The communication between client and server isn't HTTP, reducing the protocol overhead.
- The communication is bi-directional, enabling client and server to communicate over the single connection without having to open or close connections.
- We can implement fault tolerance and guarantee message delivery.

- Due to the lower level nature of the protocol, the streaming API and clients can maintain a higher level of throughput (measured in messages per second).
- This protocol allows us to develop a custom sub-protocol—the messages passed back and forth between the client and streaming API after the connection is established.

For streaming systems that use a client capable of handling HTTP, WebSockets is fast becoming one of the most-used protocols. It is the protocol to choose if you want to provide a fault-tolerant and reliable API. As mentioned, using this bi-directional protocol you can control the server and client sides of the communication. If you're building a system that uses resource-constrained clients that don't support HTTP, you may want to look at two other protocols: Data Distribution Service (DDS) and MQ Telemetry Transport (MQTT). Both were designed more for a Publish-Subscribe type model, but you may be able to use them for a stream of data.

We've covered a lot of ground talking about each of the protocols. Table 7.1 summarizes what we've talked about and can serve as a guide when you need to compare them.

Table 7.1 Summary comparisons of the different protocols

Protocol	Message frequency	Communication direction	Message latency	Efficiency	Fault tolerance/ Reliability
Webhooks	Low	Uni-directional (server to client)	Average	Low	None
HTTP Long Polling	Average	Bi-directional	Average	Average	None
Server-sent events	High	Uni-directional	Low	High	None by default. Has the ability to partially implement.
WebSockets	High	Bi-directional	Low	High	None by default. Has the ability to implement completely.

7.3   Filtering the stream

Before discussing aspects and nuances of filtering, let me define what I mean by *filtering*. Filtering in the context of the streaming API is the ability to restrict the events emitted and the properties of those events to only those of interest to a streaming client; this may be different than the filtering or reducing we learned about in our discussion of streaming algorithms. When considering filtering the stream, you need to understand the following:

- *Where the filtering is happening*—Analysis, streaming, or client tier (the client tier is our focus in the next chapter)
- *The type of filtering*—Static or dynamic

7.3.1 *Where to filter*

When it comes to deciding where to perform the filtering, you should consider several things. If any aggregations or other streaming algorithms are being applied to the stream, make sure all filtering happens in the analysis tier. Let's say we're interested in the hourly energy consumption along with the running total consumption for all homes in Chicago, IL, every 30 seconds. Given what we've learned, performing this type of computation in the analysis tier would be easy, and that would be the most appropriate place to perform that filtering. In some cases, if the volume of events/sec is low, an analysis tier may be overkill, in which case you may be able to do the filtering in the streaming API tier. Keep in mind that as the volume grows and/or the computations become more complex, you will want to look at adding an analysis tier. A use case where performing the filtering in the streaming API tier may make sense is if the stream you're producing is raw or slightly augmented, meaning the messages being emitted look the same or are close to the same as those ingested into the pipeline.

Imagine we have a stream of GPS location and vehicle metadata for all cars on highways in the United States. A streaming client may want to filter this data down to the events occurring in a certain geo-location, or by other metadata—for example, car make and model. This type of filtering would work fine in the streaming API tier. As the data is flowing through, we would filter out the events that did not match the criteria.

That brings up an important point: a user (streaming client) will be interested in both filtering whole events and in filtering out different properties from the events. If you're thinking this sounds a lot like SQL, you're correct, and that's a good way to think about it. A SQL `select` clause provides a way to retrieve only the data you want, using a `where` clause and filtering the data down to only the columns of data you are interested in. This is exactly what your users will want; conceptually the table in the relational world is now a stream, and the columns are properties on events.

The next section considers the two types of filtering.

7.3.2 *Static vs. dynamic filtering*

Filtering types fall into two buckets: static and dynamic. *Static* filtering is where the decisions about what the stream will contain are made ahead of time. You may think of this as a canned or out-of-the-box stream. It's the one that you as the streaming platform developer or architect decide on, and the client can't change the filtering being applied. From the relational standpoint this is similar to a view—the designer/developer of it decides on the data, and the user can't change it; it's the same for everyone.

Dynamic filtering is where the filtering is decided at run-time, and the streaming client can drive it. From a relational standpoint, the dynamic filtering is akin to running an arbitrary query against a table. In the streaming world, the "table" is the union of all the event schemas in the stream, giving the streaming client a lot of power. Given these different approaches, you'll also want to determine how to implement it with the protocol you chose or are considering for your streaming API.

Table 7.2 shows potential ways to think about integrating the different types of filtering for each of the protocols we've discussed.

Table 7.2   Considerations for filtering with the different protocols

Protocol	Dynamic filtering	Static filtering
Webhooks	When the endpoint is registered, you'd need a way to capture a query that can be used.	When the endpoint is registered, you'd need a way to capture a query that can be used.
HTTP Long Polling	You'd need to provide a means for the query to be expressed in a URL; the easiest thing is perhaps the URL itself representing the filter, such as '/top/50/products/viewed.'	You'd need to provide a means for the query to be expressed in a URL, perhaps using query parameters.
Server-sent events	You'd need to provide a way for the user to encode the filtering when they specify the URL for the EventSource constructor, perhaps in a similar fashion to HTTP Long Polling.	Similar to HTTP Long Polling, you'd need to provide a means for a user to specify the query as query parameters for the constructor of the EventSource.
WebSockets	This protocol provides the most flexibility. One way to handle this is after connecting, the client can send a message indicating the filtering; in the case of static, it can be a name.	One way to handle this is after connecting, the client can send a message indicating the query to be used for filtering.

As you can see, each protocol provides a way to communicate the desired filtering—the key difference is the level of flexibility you have in supporting it. When considering the design of your streaming API, remember that many developers and business people are familiar with SQL. Therefore, as you build your APIs and think through your filtering design, give some thought to how you can leverage their SQL knowledge and adopt an SQL-like syntax to your filtering. There are ways to add SQL-like query capabilities to your API. If you're building your pipeline using a JVM-based language, one option to explore is Apache Calcite (https://calcite.apache.org). Even if you're using a different language, this project may provide some ideas while you also evaluate SQL-parsing options for your chosen platform.

7.4   *Use case: building a Meetup RSVP streaming API*

Let's come back up for air and apply what we've learned to a use case. Let's pretend that we want to build a streaming system that uses the Meetup.com RSVP data as the data source. You can see this data source in action at http://stream.meetup.com/2/rsvps. At this URL you see a constant stream of JSON data; each entry is the result of someone entering an RSVP for a Meetup.

From this data source, imagine that we want to allow users to connect to our streaming API and filter the data by topic name. Because we're talking about the streaming API in this chapter, let's focus on making the data available to consumers.

In the second part of this book, when we build a real solution, we'll have to address how to consume the data from an analysis store. For now, we'll only worry about the communication and protocol choices we need to make. Figure 7.11 gives a depiction of what we're trying to accomplish.

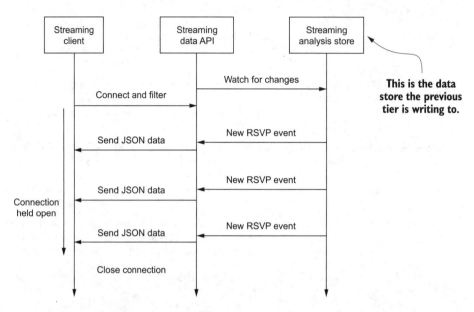

Figure 7.11   Meetup RSVP streaming API data flow with filtering when connecting

Figure 7.11 starts with a client connecting to our streaming API and passing in a filter. Our streaming API watches the analysis store for new data and then sends all new "RSVP Events" to the client. The two questions we need to answer are which communication pattern to use and which protocol to use? Tables 7.3 and 7.4 outline the communication and protocol choices, along with their applicability to this use case.

Table 7.3   Which communication pattern to use?

Communication pattern	Good fit	Comments
Data Sync	No	In this use case we're not transferring a complete data set, and the fact the data is continuously flowing would make the Data Sync protocol a bad choice.
RMI/RPC	No	Using RMI/RPC would not be a good fit here because we're looking at the data flowing toward the consumer and not having to call a method on the consumer each time a new event arrives.

Table 7.3 Which communication pattern to use? *(continued)*

Communication pattern	Good fit	Comments
Simple Messaging	Maybe	This pattern would not be a great fit because we're trying to do more than return the most recent data. In this case we're also applying filters to it. Although it may be something that can be made to work, there will be a lot of potential overhead on the API server to implement it.
Publish-Subscribe	Yes	This would be a good choice for this use case. We could gain some efficiency with the different protocols that can be used with this pattern, and the model fits how the data is flowing.

It looks like we have two communication patterns that would be candidates for this use case. Now let's look at the protocols.

Table 7.4 Which protocol to use?

Protocol	Good fit	Comments
Webhooks	No	There would be a lot of overhead because the streaming API would need to make an HTTP POST request to the registered callback for every new event. Not a good fit.
HTTP Long Polling	Yes	This protocol would work fine in this use case because our filtering is being sent during the client connection. To make that work with this protocol, we would want to allow the client to use a query parameter on the URL to indicate how they want to filter the results.
Server-sent events	Yes	This protocol would work well, given the same caveat as with HTTP Long Polling, that we're allowing the user to pass in the filter when they connect via a query parameter.
WebSockets	Yes	This protocol is a good choice as well. As with the other protocols that are good for this use case, we would need to have a way for the client to pass in the filter criteria. Where this protocol differs is that it can be a query parameter or a message that is sent when the client initiates the connection.

Looking at the communication and protocol options for this use case, we have a number of options that would allow us to build a streaming API that meets our needs. You should take several things into consideration when deciding which communication pattern and protocols to support. First, consider the clients you want to support and how you want the filtering to be performed. To allow the largest number of clients, providing Long Polling, server-sent events, and WebSockets would be the way to go. But if what is more important to you is that you allow filtering after the connection is established, then you need to use a more robust protocol such as WebSockets.

7.5   *Summary*

In this chapter we looked at the communications patterns and protocols used for sending data to a streaming client. We also talked about filtering a stream of data. We didn't dive too deeply into the different ways of implementing fault tolerance because they are all similar to what you've seen in the other chapters leading up to this.

Key take-aways include:

- It's important to closely look at your requirements and match up the different communication patterns and protocols. There is no one-size-fits-all prescription.
- When choosing the communication pattern and protocol, pay particular attention to your needs for fault tolerance and reliability, because they will drive your choices.
- Don't overlook supporting static and/or dynamic filtering. You may not think about them at first, but filtering is a feature your clients will want soon after your API is available.

Some of the protocols may not cover exactly what you need for your business problem, but I hope at least that after this chapter, you feel comfortable looking at other protocols and have a better understanding of how to approach looking at a new protocol.

Consumer device capabilities and limitations accessing the data

8

This chapter covers

- Core principles of a streaming client
- An introduction to the web client
- The quest for querying a stream

Believe it or not, we've almost completed our journey from initially receiving the data in our collection tier, through moving it, analyzing it, and making it available for consumption. Now we'll finally take action on the data. Figure 8.1 shows the overall architecture we've been discussing, with this chapter in focus.

For many, the technologies covered in this chapter, along with the ideas and myriad solutions, are where a streaming system comes to life. Up to this point, we've been dealing with what many would call back-end technologies—your end-users or customers rarely if ever see all the work that goes in. This chapter focuses on the last mile, if you will: taking the data from the streaming API to the client. We will be talking about taking action on the stream of data; this may be with a mobile application, web browser, vending machine, farmer's combine, another system in an enterprise, or any connected device.

The things you can do in this layer are wide open and exciting; they can be almost anything you can think of. Perhaps you want to build a dashboard showing

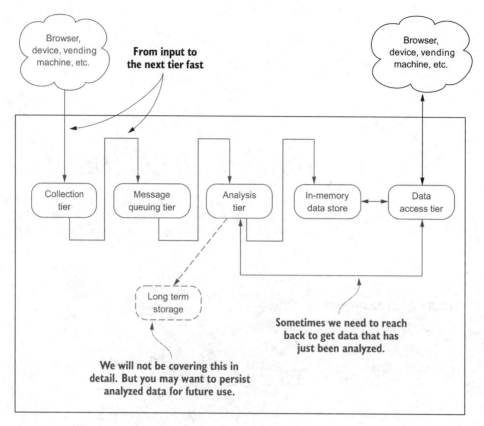

Figure 8.1   Architectural blueprint with this chapter in focus

what's happening right now in your business, offer real-time product recommenda-
tions to your customers, provide real-time analysis and feedback on a golf swing, or
provide real-time adjustments to the route and speed of a locomotive so it can conserve fuel. The concepts and technologies that make all this possible are the focus for
this chapter. In the intro to chapter 7 I said that it was my favorite to build. As a
streaming data geek, that may be true. But the products and results that come out of
this tier turn the lights on for people, show the power of building a streaming system,
and bring all the hard work to life.

It's tempting to consider covering all the different clients out there today. The
landscape is changing so fast, however, that providing such coverage at any reasonable
level is impractical. What's important is understanding the core concepts that apply
today and that will continue to be relevant even as the client devices evolve at a break-
neck pace. We will stick to the concepts, then, and not discuss the technical features of
all the devices on the market. Then we'll work through examples of clients that
embody these concepts.

Without further ado, let's refill our coffee cups and jump in.

8.1 The core concepts

Myriad products you can build in this tier all fall into one of three categories illustrated in figure 8.2.

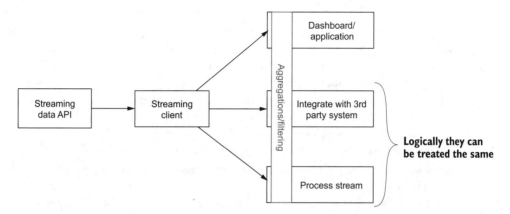

Figure 8.2 **The three general categories of client applications**

Figure 8.2 doesn't call out the particular protocol or communication pattern being used between the streaming data API and the streaming client. The protocol will be important when we discuss the core concepts in more detail, but let's not worry about that now. You can see aggregations and filtering crossing all the application categories. I consider these crosscutting concerns—regardless of the client application you're building, there will be times when you want to aggregate and filter the stream. Sometimes that will happen client-side, but many times you'll want the streaming system to do it. We'll return to this balance in more detail later.

First I'll elaborate briefly on each of the categories from figure 8.2 before explaining further.

UI/END-USER APPLICATION

The central theme of this category (labeled "Dashboard/Application" in figure 8.2) is that it's an application—or more realistically a component of an application—that consumes a stream as part of its normal processing. Applications in this class cover a wide spectrum, from simple web browser–based dashboards (okay, admittedly these can be quite complex), to social chat or messaging applications running on devices, to even multiplayer games. In general, these are applications that connect directly to a stream and handle all manipulation of data on the client side.

INTEGRATION WITH THIRD-PARTY/STREAM PROCESSORS

We can group integration with a third-party application and a stream processor logically into one category. All these applications follow the same general pattern: they consume a stream, perform some business logic, and then interact with another system. That other system may be external to your organization or may be another part of a larger application completely under your control. For example, perhaps you're consuming a stream of transaction data that you're sending to an external fraud-detection system for analysis. You may process some of this data, but in general you're consuming a stream and sending the data to a system that's not under your control. Another example of interacting with a third party would be building a system that ingests a stream of customer behavior data—perhaps marketing emails that are opened or ad impressions viewed online—and then computes a propensity-to-buy score that's then updated in SalesForce.com.

Many times you may build a stream-processing application that only interacts with other systems internal to your organization. For example, you build a distributed system composed of small services that all need to communicate in a workflow. One way to do that would be to set up the services as producers and consumers of a stream. They would each "listen" to the stream for work to do, perform the work, and then send the results of their work back into the stream to be picked up by a different service in the workflow.

The take-away for these two categories isn't to provide a cut-and-dry bucket for your streaming client application to fall into but to give us a frame of reference as we walk through the different core concepts. With that in mind, let's begin our journey through the core concepts. For each concept, we will talk about the strategies when consuming from a third-party API and the strategies when it is your own API. The strategies when it's your own may lead to internal discussions about how to change the streaming API discussed in chapter 7, and that's okay. There should be a healthy tension between the two layers of the architecture, and it will require discussion between teams that are building each and an understanding of the needs at each layer.

All right, let's get going and talk about the concepts.

8.1.1   *Reading fast enough*

Speed of reading may not be the first thing you think of when thinking about building a streaming client, but it's often a major consideration. There are two important sides to ensuring a client is reading fast enough: the streaming API side and the streaming client side. This chapter is focused on the streaming client, but let's look at the streaming API side of the problem for a moment.

Why is reading fast enough important to the API? It boils down to data loss (first and perhaps foremost) and server resource utilization. This will depend on the technology used to deliver the stream from the API. In our case, let's zoom in on two popular techniques discussed in chapter 7: server-sent events and WebSockets, which are similar from both a server and client side. Figure 8.3 shows the client-not-reading-fast-enough situation we can run into with both approaches.

In figure 8.3, the client is able to process the first two messages before another message is ready to be sent. But processing message 3 took too long, resulting in the streaming API having to decide what to do with messages 4–6. Should it hold them in memory or blindly send them? Holding in memory is certainly an option in this

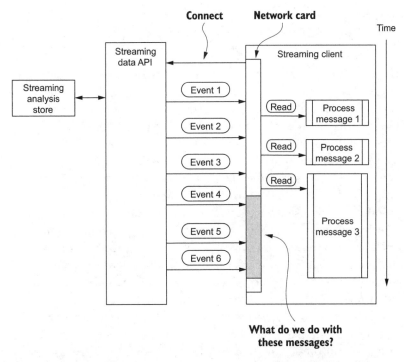

Figure 8.3 Generalized server-sent events and WebSockets data flow showing slow client

example—we're only talking about three messages. Blindly sending may also work, although we would have to deeply understand the underlying technology the streaming API is using. What if that technology drops the messages if the network buffers are full? In this case, there is the potential for data loss. Let's take the approach that the streaming API will hold the messages in memory.

This may seem like an easy problem when we're talking about a couple of messages. What about when the velocity of the stream is 1000 messages/sec, and the client is falling behind? The API developer is left in a position of potentially having to discard data to ensure that server resources aren't exhausted or applying backpressure to upstream systems. To aid in this situation and help to ensure data isn't lost and the server resources aren't exhausted, an API developer can notify a client that it's falling behind. Unfortunately, not all third-party APIs offer this feature, but if you're developing the streaming API, I highly recommend that you do provide this to your clients.

Turning our attention back to the client side of this problem, three important questions pop into my mind:

- How do I know if I'm reading fast enough?
- What happens if I'm not?
- How do I scale my client so I can keep up with the pace of the stream?

The way you address these questions will vary slightly depending on whether you're consuming a third-party API or an internally developed one.

THIRD-PARTY STREAMING API

Depending on the streaming API you're connecting to, it may provide guidance on how to be notified if you fall behind and what the ramifications are for doing so. For example, as of this writing, the Twitter API will disconnect any client that falls too far behind. Twitter doesn't provide an explanation for what "too far behind" means, but it does send *stall warnings* that your client is falling behind. The Twitter API sends a message approximately every five minutes if a client is falling behind. Again, there's no guidance on how many stall warnings will be sent before your client is disconnected.

What can you do if the API you're consuming data from doesn't offer a warning message or other mechanism to let you know you're falling behind? One strategy is to read the timestamps on the messages you're receiving and compare them to the current time. Then as you process messages, if the gap between the current time and the message time starts to drift you can reason that your client software may be falling behind. Remember that the service you're consuming from provides a timestamp on each of the messages being sent. If the stream you're consuming doesn't provide a timestamp on each message for when it was generated, your next best bet would be to try to ascertain what the expected flow rate is for the stream. Some streams may be episodic in nature; in these cases you may be able to ascertain the pattern and from there reason about how well your client is keeping up with the stream.

This brings us to the next question: What happens if we are not keeping up? This is an interesting question, and unfortunately the answer depends on the streaming API you're consuming data from. As of this writing the Twitter API clearly indicates that it will close the connection if your consumer can't keep up with the stream. Other streaming APIs may not drop the connection; in fact, some will instead drop data that can't be consumed fast enough. If you're building a consumer that consumes data from a third-party streaming API, make sure you ask about this scenario.

YOUR STREAMING API

If you're building a solution whereby you control the entire stack, then you want to ensure your streaming API clearly states what happens to the data and/or the connection if the consumer can't keep up. You can do that via documentation, including status messages in the stream and logging the sending of status messages. The status messages should provide information on how far behind the consumer is and a warning about the connection being closed if the consumer doesn't keep up. We all know that sometimes documentation doesn't get read, and having a data-driven solution will allow your customers to react to changes and allow your streaming API to provide a better experience. Be aware that this will put a further burden on a client that's already having trouble keeping up, by asking it to process another message type and act on it. You should also log the sending of these messages so that they can be analyzed to help in troubleshooting slow consumers and/or problems in your streaming API.

8.1.2 *Maintaining state*

Handling state in a client-side streaming application can come in many forms. First, you may want to preserve the state of where you left off, perhaps storing metadata about the last message so that when the user decides to start the application again, you can begin by consuming from the last message processed. You may also want to store the updated results of a streaming computation that is performed client-side. This would enable you to regularly update a UI with the up-to-date results as you're processing new data in the background. These are great reasons to maintain a working state in the client; it's applicable to many use cases and should have a negligible impact on the client.

Be careful not to take this too far, resulting in your client performing streaming computations as data is flowing on the client side. Certainly, some computations make sense. I would draw the line at doing any window-based computation client-side. Even with this line in the sand, it may still seem like a good idea when the data flow has low volume and velocity. Invariably the data stream will grow over time—otherwise you wouldn't invest the effort in building a streaming system. You can probably think of streams of data that you don't want to grow. For example, nobody wants the stream of error conditions encountered in a manufacturing plant to increase in volume or velocity. But be careful and don't fall prey to saying, "Oh, for this one stream we'll do client-side computations." Once you start down that path, you'll end up with a client-side code base that is fragile, inconsistent, and—worse—will fail when the stream does grow. That said, it all sounds great on paper to say we're going to push back and *not* do client-side aggregations or computations. The reality is that there *will* be times when you need to maintain state or do client-side computations when building the client side of a streaming application. Let's walk through options on how you can go about doing this.

Given the sheer number of options to implement client software, we're going to illustrate only how to handle maintaining state using browser-based clients. Even though this leaves out various client devices, the general principles will still apply. Regardless of the technology you use for your client, for high-velocity and volume streams you need to pay special attention to how you structure the data and the memory that's consumed. I've seen streaming clients implemented in browsers that over time will run a browser out of memory and crash when performing computations over windows of time that should have been performed by the streaming analytics platform.

When building a web-based streaming client, there are two main options for storing state: web storage and IndexedDB. Web storage is further subdivided into local storage and session storage. The web storage options are designed to store key-value pairs in a more intuitive way than in cookies. Local storage is designed to allow data to be persisted between when the browser is closed and reopened. Session storage is subdivided by the domains that you store data for and only survives a browser session.

There are limitations with the web storage options. First, the order in which the keys are stored is user-agent dependent. Therefore, you can't count on accessing data from a stream in the same order you received it. But if you're storing the top 10 events, you can retrieve them all and sort them in memory. The second limitation is that the web storage options are limited to 5 MB per origin. This isn't a hard-and-fast rule, but it is the recommended limit in the specification. To learn more about this storage option, see the W3C specification at http://www.w3.org/TR/webstorage/.

Indexed/DB, the second storage option, picks up where web storage leaves off. This one is designed for in-order retrieval of keys, efficient searching over the values, storage of duplicate values for keys, and the in-order traversal over large numbers of records. To learn more, see http://www.w3.org/TR/IndexedDB/.

When it comes to keeping state client-side with a browser, either storage technology should suffice, as the state should be small. But if you find yourself in the situation where you need to do computations and/or aggregations client side, then I recommend using IndexedDB as the storage mechanism. Again, if possible you should push back and try to have this work performed behind the streaming API. If you're also building the streaming API, you should spend time thinking through how you can perform the required computations for your clients as well as offer a means to offload any state management from them. The goal should be to make the client as stateless as possible. You may think you're going to use Java on the client side or Node.js, so you don't have to worry about this. You may not have to contend with browser technologies, but you should consider that at high-speed you need to make sure you're not keeping too much state or trying to do advanced computations/aggregations in your client. At a certain data rate, you'll exhaust the resources of the client environment. That's why, regardless of client technology, I'm still a big fan of trying to push back to the streaming API team to have the computations performed in that layer and work toward keeping the client stateless.

8.1.3 *Mitigating data loss*

When it comes to mitigating data loss, there is a client-side aspect to be concerned with. Figure 8.4 identifies different areas where we can have data loss.

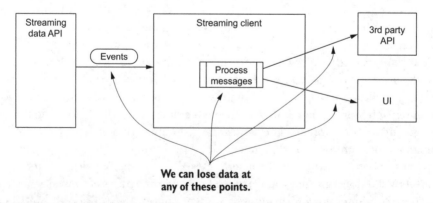

**We can lose data at
any of these points.**

Figure 8.4   A streaming client with potential areas of data loss in receiving, processing, and sending data

Figure 8.4 shows four main places where we may lose data. First, when it's sent from the streaming API; secondly, when we're processing it; and finally, when we're sending it to a UI or a third-party API. We have a couple of options for mitigating the loss of data in each spot. Our discussion will focus on receiving and sending, not on processing, because data loss while processing it is almost entirely a result of a programming bug. And we'll eliminate the UI from this discussion because it's completely within your control, leaving the receiving of data and sending of data after we process it the areas to address.

We'll focus on a non-web-based client. Imagine that we're consuming a stream of data that contains order transactions and need to send those to an order entry system. This type of integration is often done with client software that is not a web browser. In this case, we want to make sure that we don't lose any data coming from the streaming API and that the third-party system received the new message. To do this, we'll use concepts we talked about in chapter 2: receiver-based message logging (RBML) and sender-based message logging (SBML). When used together RBML and SBML are commonly implemented using the hybrid message logging (HML) pattern discussed in section 2.3.3. Let's apply those patterns to this problem and see if we can reduce our likelihood of data loss. Figure 8.5 shows where we would add in the receiver and sender aspects of the HML solution.

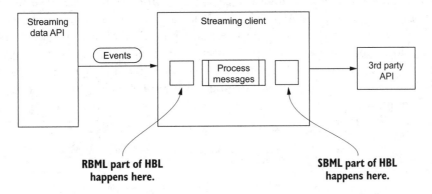

Figure 8.5   HML in perspective to mitigate data loss

The general flow for how this would work goes like this: first, we implement the receiver side of HML right after we receive the message from the streaming API. We then process it and before sending it to the third party we implement the sender part of HML. That's pretty straightforward, but how do we know when to delete the messages from both of the logs? For the receiver side, once we're done successfully processing the message and have recorded it in the sender-based log, we can remove it. But how do we know when to delete the message from the sender-based log? To do this we need to rely on one of two methods: acknowledgment and the write-and-read pattern. If the third-party API returns a success code to us upon successfully receiving the data, we can use that as an acknowledgment and remove the log entry. This is the best we can hope for—it's architecturally clean and a natural way to program the client.

Unfortunately, not all third-party APIs supply that type of request response. Our next best bet is the write-and-read pattern, where we send a message and then try to query the system afterward and see if we can retrieve the data we sent. This is not a good pattern, but it is an approach to work around poorly designed software. It can get messy the more APIs you're integrating with and can become fragile, although sometimes it's the best you can do. Even with these two strategies there will still be situations where neither option is a possibility. In that case you need to get creative in how you ensure that the data was received and the log entry can be removed.

Looking again at the receiver side, where we get the data from the streaming API, is implementing the receiver side of HML enough to ensure that we've fully mitigated the potential for data loss in the use case where we're mitigating the loss *after* we receive the data? I would say yes, it is.

8.1.4   *Exactly-once processing*

There are plenty of uses cases where being able to process a message exactly once is important. In the case from the previous section, we're receiving data from an order stream and need to update a third-party order system. Because we didn't implement the third-party API, we can't make any assumptions regarding its idempotency. Therefore, we need to ensure that we don't send it duplicate messages. Ideally, to handle this we would be able to use an acknowledgment feature of a streaming API. This interaction pattern—acknowledgment from the third-party API and the streaming client acknowledging the message being handled—are illustrated in figure 8.6. Again, due to the nature of this work, using a web browser as a client is impractical.

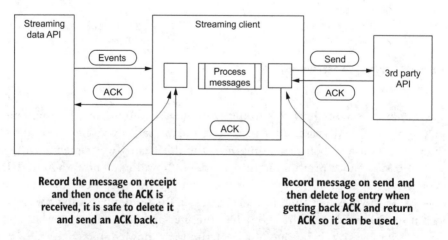

Figure 8.6   HML with full acknowledgment shown in context

This is fraught with challenges. First, you need to implement message acknowledgment for the third-party API as discussed in chapter 7. Then you need to implement it with the streaming API. Unfortunately, in many cases you don't control the streaming API and may have to deal with the inability to send back acknowledgments. How can you ensure that you've processed a message exactly once? In this case we need to keep a record of all the messages we've seen for some time period. If there is a guaranteed unique message ID, we can use it in our "been processed message store"—otherwise the data can be hashed or perhaps another fingerprinting mechanism can be computed. By keeping a record of all the messages we've seen, we increase our client complexity and storage requirements but we also protect ourselves from when the streaming API crashes. The architecture for this looks like figure 8.7.

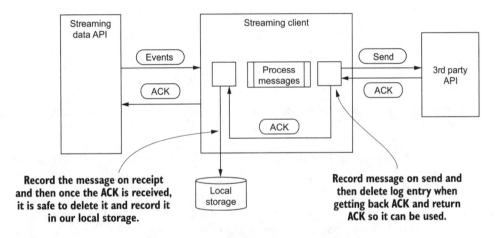

Figure 8.7   HML with partial ACKing and local storage

This may seem simple enough, but what if there's more than one streaming API server? At a certain point, one is going to crash or be taken down for maintenance. When this happens our streaming client will connect to a different API server and begin to consume the stream. How do we ensure that we don't process a previously processed message again? Remember, we can't assume that the streaming API is keeping track of what messages were already delivered to us. Perhaps it does during normal maintenance, but what if it crashed before it could record the last message sent to us and sends us the same message again? To adequately handle this, we'll need to use a distributed store for recording the previously seen message. This is illustrated in figure 8.8, where there is more than one streaming API and a streaming client.

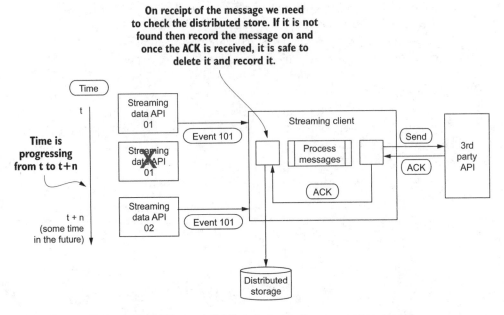

Figure 8.8   HML with partial ACKing and distributed storage for processed messages

To keep things simple in figure 8.8 I only show one streaming client, but I think you get the picture. As time goes on we see that Streaming API Server 01 goes away and Streaming API Server 02 sends our streaming client a message we've already seen. In this case and in all cases we need to check whether we've already processed the message, and if so we need to discard it. There is a subtlety to this design that you need to keep in mind. We're using a distributed store for keeping track of previously processed messages. Fantastic—now we've protected ourselves from double-processing messages. But if we're processing a stream that has a significant velocity, any requests we make to a service that is over the network may potentially jeopardize our ability to meet an SLA. Keep this in mind, measure the performance, and if you notice an impact consider using a local store that flushes its messages to a distributed store. Figure 8.9 illustrates how this may work.

When the streaming client detects that the network connection to the streaming API server has been lost, it needs to synchronize the local store with the distributed store. As long as this step happens prior to receiving another message to process, then we should be able to ensure that we don't process a message more than once.

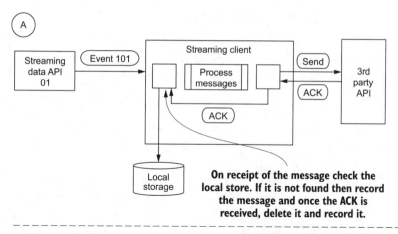

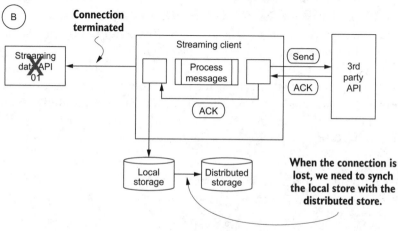

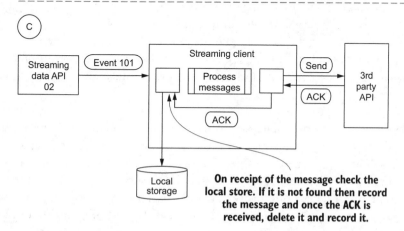

Figure 8.9   Time lapse of handling a streaming API server changing and keeping exactly-once processing

8.2   *Making it real: SuperMediaMarkets*

Phew, we made it through a lot of material! Most of it is low-level detail regarding things to keep in mind and sometimes do when building a streaming client. You may wonder how/why/when would you use all this? Those are great questions, in particular if you've been thinking of using a browser as a client or have seen an example of a web client consuming a stream. I totally understand—in fact the next section goes through a web client use case. Before proceeding to that, though, I want to stop here and go over a concrete example of a use case where you'd need to think through and apply the concepts we've been talking about.

Imagine we work for a company called SuperMediaMarkets. Our business is providing an online marketplace for media assets, playing middleman between the producers of media assets (photographers, video producers, musicians, artists, and so on) and consumers (marketers, web designers, bloggers, and others). The way it works is that a media producer uploads an asset, we sell it to a consumer, and for our hard work we take a percentage of the sale. You can think of it like eBay for digital media. As a popular service, we have people buying and selling assets on our site all day long. Recently a team at our company built a streaming-data pipeline that captures everything happening on our site and makes it available to other developers to build streaming applications. Now that this stream is available, we would like to build a client that would let us process sales, computing and potentially paying royalties in real time. For our real-time royalties client to be successful, we need to meet the following requirements:

- We must only process a royalty exactly once
- We must not lose a royalty

Our streaming team provides a WebSocket API for us to use, so we've decided to use Node.js to implement our streaming client. Not to worry—we're not going to get into the weeds on exactly how you would do this in Node.js. If Node.js *is* your cup of tea, you should be able to use the following discussion as a guide to implementing a robust streaming client. With that said, let's peel back the layers of these requirements and discuss what we need to build and what we need from our streaming API to accomplish this.

How are we going to meet our first requirement of "only process a royalty once"? This is an interesting requirement. Up until now we've been talking about our streaming pipeline, something totally under our control, having exactly-once semantics. But our client is going to sit between our streaming API and a third-party processing system. To meet our requirements goal we need to make sure that we (a) only send a payment request one time to our processor and (b) make sure we only receive that message one time from our streaming API. Figure 8.10 shows the data flow of how we can do this.

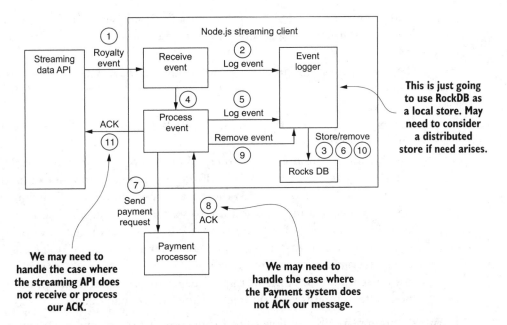

Figure 8.10 Node.js client with the steps required for exactly-once processing

A lot is going on in figure 8.10. Let's discuss some of the issues we may still need to deal with that aren't handled in the figure. The steps in the flow are as follows:

1 *Royalty event*—In this step we receive the event from the streaming API.

2 *Log event*—This is the RBML side of the HML algorithm. Once we receive the message we store it.

3 *Store event*—We store the event in RocksDB which works great as a local store that meets the needs of fast insert and delete.

4 *Process event*—At this point we've stored the original event we received and can now send it safely.

5 *Log event*—We will log the event before sending it—this is the SBML side of the HML algorithm, where we store the message we are about to send in case we have a problem during the send.

6 *Store event*—Store the event as we did in step 3.

7 *Send payment request*—We're now ready to send the payment request to our payment processor.

8 *Receive ACK*—Wait until we receive the "ACK" that our payment request was handled successfully. We may have to do something different here in the event our payment processor doesn't support ACKing. In that case, you may want to check to see if there's a way you can query the system to confirm that your request was successful. Keep in mind that in some cases this may not be possible,

and you may need a different solution that is more asynchronous and waits longer or that has to run a separate process to determine if the payment was processed successfully.

9 *Remove event*—At this point we know that the payment system has processed the request, so we can safely delete the royalty event received in step 1 from our backing store.

10 *Remove*—This is a remove from the RocksDB. In this case, it will remove the royalty event we received in step 1.

11   *ACK event*—At this point we've finished processing the event and need to ACK the message to our streaming API. If the streaming API doesn't support ACKing, there will have to be some other means by which you can let it know you're done processing a message and not to send it to you again. There's a chance that the streaming API doesn't receive our ACK due to network problems or doesn't process it. In that case we can't guarantee that we don't receive the same message a second time. Therefore, we may need to add a check in step 7 before sending the payment request to the processor to help ensure that we don't attempt to process a payment request more than once.

With all this in place, we can now safely handle our first requirement. Do you think we can satisfy the second requirement, to not lose data? In general, we could handle this requirement without any changes. We must be cautious of not doing too much work in the box that shows we are receiving the data. We want to write it to a durable store as soon as possible—it's during this time that we're at most risk for losing data. After we make changes to it, we also write it to a durable store.

There is still risk that we can lose data. What happens if we lose the disk that RocksDB is stored on? We would lose the data. We have several options to handle this, but I'll mention only two. First, we can use a RAID configuration so we have storage redundancy. Second, we can choose to use a distributed store. With the distributed store, you need to ensure that the data is received and stored, which adds more complexity. Some of these extra steps could be avoided if the streaming pipeline that another team built has a way for you to ask for data that occurred in the past. In that case you need to decide how much you want to put the fate of data loss in someone else's hands.

Hopefully you now have a better idea of how to apply the concepts discussed earlier in the chapter. Next we are going to look at the web client and how to provide a dashboard for an end user.

8.3 *Introducing the web client*

Let's take a break from low-level details and talk about using a web client as a streaming client. Using a web client, we no longer need to think about a lot of those details because the web client isn't used to interact with a third-party API that may be transactional and is perhaps not idempotent. Instead the web client is primarily used for displaying the results of a stream. We will often need to keep state and potentially perform some computations. Although there is and will continue to be a tremendous amount of growth in the industry around building clients that act on data, being able to use a browser to display the results of a streaming application is today and will for quite some time be a useful tool because it allows people to understand their business. Some may question the use of a browser for rendering a stream, calling it "eye-candy." At times I can agree with that, although recently I was working with two clients that used a browser-based dashboard for the following two reasons:

- It was the first time they could understand how people used their website and understand how their business was doing. You may think you can get this from a typical analytics report, but when you see the stream of data, you see that it's alive and lets you see a different story unfold.
- A major news publisher used a browser-based dashboard in its editorial bull-pen so editors could see in real time how their articles were performing. They could use the tool to help make decisions on what types of social promoting to try and drive ad revenue for their articles. This is powerful and may be hard to automate.

If you have a stream flowing at a high velocity, a browser-based client will undoubtedly fall behind in reading it. That may sound like a terrible thing, but I equate it to deciding which HD TV to buy. When I compare the picture of a lower-end HD TV to a high-end HD TV, I may notice subtle differences in the color saturation or sharpness, but the picture doesn't change. This same idea applies to a dashboard in a browser that can't keep up with a stream. It may be missing some of the data points, but the overall picture of the business and trends doesn't change. In working with hundreds of customers I've seen this concept proved out over and over. Everyone says they want absolute 100% accuracy in the numbers, but when it comes down to it, being able to tell the correct story and show the correct trends is more than satisfactory.

That doesn't mean we don't need to worry about anything. We will undoubtedly have to keep state and may have to perform some basic computations. Earlier I talked about maintaining state using web storage and IndexedDB in a browser. Without getting too bogged down by the code, let's see how we would want to use those technologies to store state and perform basic computations. Figure 8.11 illustrates a web-based client that's showing a stacked bar graph of running sales per rep.

In figure 8.11, as the data is flowing from the streaming API, we'll be using JavaScript in the browser to update the in-browser local storage. There are two options when storing the current state of the graph in local storage: store the latest values received from the streaming API or the current state of the entire graph. The benefit to storing the entire state of the graph is that if the user closes the page and comes back to it, we have some state to start off with, and if we use local storage and not the session storage, then the state will persist after the user closes the browser.

There are two things to keep in mind with this. First, before deciding to store the data in local storage and having it persist after the browser is shut down, you will want

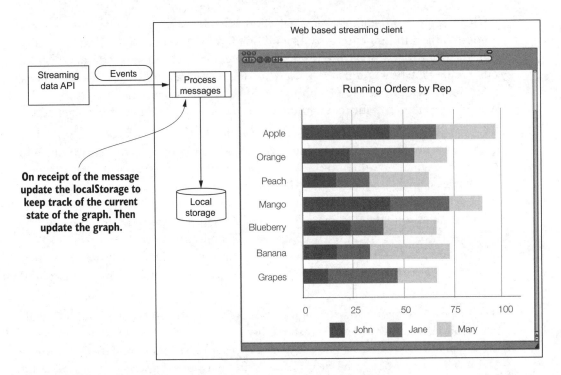

Figure 8.11   Web-based client using local storage to maintain state of a graph

to check with the appropriate security personnel to ensure it's appropriate. Secondly, there may be many times when the streaming API is not sending the running totals but only the incremental changes. When this occurs you want to be able to keep track of the changes and increment the values accordingly. If you find you have to do this for a lot of different data points, consider using the IndexedDB storage option and not local storage.

8.3.1   Integrating with the streaming API service

When it comes to integrating a browser-based client with a streaming API there's a lot of flexibility. Out of the box you should be able to integrate with APIs that expose a stream using HTTP Long Polling, WebSockets, and server-sent events. The next listing and listings 8.2 and 8.3 illustrate these technologies.

Listing 8.1   Server-Sent Events

```
var source = new EventSource("<SSE Service URL>");
```
Connects to the server-sent event server

```
source.onmessage = function(event) {
    document.getElementById("event").innerHTML += event.data + <br>";
};
```
The function called every time a new message is received

Listing 8.2 HTTP Long Polling

The function called
every time the
timer gets called

The AJAX function with
the URL we need
to communicate to

```
(function pollServer() {
  setTimeout(function() {
    $.ajax({ url: "<Server To Talk To>", success: function(data) {
      document.getElementById("event").innerHTML += data.value + <br>";
    }, dataType: "json", complete: pollServer });
  }, 30000);
})();
```

The resetting of the
timer with the
function to call

The body of the
success function that
gets the HTML element
"event" and updates it

Listing 8.3 WebSockets

Establishes a
connection to the
web socket server

The function called
every time a new
message is received

```
function webSocket() {
  if ("WebSocket" in window) {
    // open a web socket
    var ws = new WebSocket("ws://<url to service>");

    ws.onmessage = function (event) {
      document.getElementById("event").innerHTML += event.data + <br>";
    };
  }
}
```

Updates the "event"
element on the web page
with the new data received

Listings 8.1 through 8.3 should give you a sense for how each of the integrations would work. Both server-sent events and WebSockets work on a message-by-message basis, allowing you to easily and cleanly apply your business logic. Arguably, the Long Polling does as well, but remember, with the Long Polling example it's the browser that's initiating the request by setting a timeout to execute the request each time.

Regardless of which technology is used by a streaming API, you should be able to integrate with it using JavaScript and a browser. Chapter 9 walks through this in more detail, and we'll build out a client application.

8.4 The move toward a query language

See chapter 9 figures 9.6 and 9.7

Shortly after you have your integration working, whether you're building a web-based streaming client or a client in your favorite programing language, your users are going to ask for the ability to apply filters and other criteria to the data stream. They'll want to be able to query it like they are used to querying traditional data stores. This may not happen at first— they will likely be ecstatic to have a stream of data flowing and to learn things about their business that they never knew. Quickly, though, the novelty will wear off, and the requests will come in. You'll be in good company because this is where streaming is going, and a lot of work is being done in the industry in this direction. Unfortunately, today you may have to do some of this client-side; even though all the stream-processing engines have or are adding SQL-like capabilities to them, the SQL support in third-party streaming APIs will vary wildly.

The introduction of SQL to a streaming client is interesting. If you've used other business systems, you've certainly used SQL. It's one of the most ubiquitous data access models there is. If the streaming API you're using provides SQL or support for a different query language, then for your client you may be able to get away with providing a thin layer over that. In essence, you offer the same support as the API and pass the query down to the API. But if the streaming API you're using doesn't support any query language, and you need to provide SQL or other query language support, there's only a certain amount you can do in the client. To properly provide query support you'll want to either involve the team that is building the streaming API you're using or, if it's a third party, consider building your own API that is a proxy to that one. Before going into the details as to why, look at the architecture illustrated in figure 8.12.

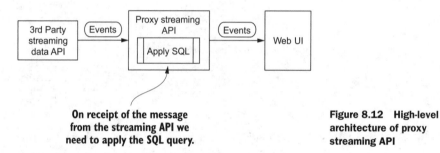

On receipt of the message
from the streaming API we
need to apply the SQL query.

Figure 8.12   High-level
architecture of proxy
streaming API

With this architecture in place we can provide complex query capabilities in our browser-based application. All we need to do is to communicate with the proxy streaming API as if it were the real streaming API. Then in the browser we can expose a way for a user to enter a SQL query, pass it back to the proxy, and have the proxy execute the query as the stream is flowing.

If you find yourself in this situation and you need to implement the streaming API proxy to provide SQL query capabilities to your users, consider using Apache Calcite (https://calcite.apache.org) for this purpose. It's a library specifically designed for processing SQL with data in a stream or at rest. There may be a Calcite adapter for your data source, but if not it's designed with extensibility in mind and should be easy to adapt to your needs.

8.5   *Summary*

In this chapter we looked at concepts to keep in mind when building a streaming client. We covered quite a bit of material that probably went deeper than you were expecting for a chapter on the client side of a stream. Some of the concepts covered here were seen before in chapters leading up to this.

The chapter covered the following:

- The core concepts to think about when delivering a stream of data
- The quest to query a stream

The key take-away from this chapter is that there is a lot more to think about when delivering streaming to users than a flashy dashboard. At a certain point, users will want to query it, and many other times you'll want to deliver the stream to a third-party application. At those points, the art of delivering a stream to a client takes on a whole new life of its own.

Chapter 9 puts this all into practice—we'll be building a complete streaming system.

Part 2

Taking it real world

T his part of the book is where we put everything from chapters 1 through 8 into practice. Chapter 9—the only chapter in this section—walks you through building an entire streaming data system. Our data set for this chapter is the Meetup.com RSVP API. We start off with a discussion of the data set and the application we're going to build. Then we will work our way through each of the tiers. When you reach the end of chapter 9, you will have a complete streaming system running. You'll have good insight into the next steps to take to bring a streaming system into production.

Analyzing Meetup RSVPs in real time

This chapter covers

- Building a complete streaming data pipeline
- Planning to take it to production

Congratulations! You have reached the chapter where we are going to take all the material you have read and put it to use. In this chapter we will build a complete streaming data pipeline and an application that consumes the stream. Instead of using a fictitious data set (and leaving you wondering how this works in the wild) we'll use a live data set—the Meetup (www.meetup.com/meetup_api/docs/stream/ 2/rsvps/#websockets) Streaming RSVP API—as the data source for our pipeline. The web application we build at the end of the chapter will allow us to glean insight from the RSVP stream. To aid in debugging, and in case the data source is no longer available, along with the code for this chapter you'll find a sample data file that you can use to simulate the stream of data. After you complete this chapter you will have a fully functional streaming pipeline and web application. With it you'll be in a good position to take it to the next level with this data set or a totally different data set.

Before we embark on our journey of building, I want to talk about two things. It would be impossible to cover every incarnation of technical choices made along the way. It is implausible to implement in one chapter everything we have discussed in the previous eight chapters.

As we implement each tier I will point out the choices you have, and I've made several choices for the implementation in this chapter:

- All technologies chosen are open source and, where possible, Apache Projects.
- Apache Maven 3.3.9 was used for all builds.
- Java 1.8 was used for the implementation language.
- The streaming application is web-based and uses JavaScript.
- The code in the snippets in this chapter is often abbreviated to call out the pertinent pieces. However, the full source is available and works on Linux, OS X, and Windows.
- The figures show an OS X terminal window, but comparable commands all work on Windows as well.

To make the download and use of the software easier, all code covered in this chapter is also available at the GitHub repository https://github.com/apsaltis/StreamingData-Book-Examples.

Figure 9.1 shows our familiar overall architecture with a small twist: this time it shows the technologies we are going to use in this chapter.

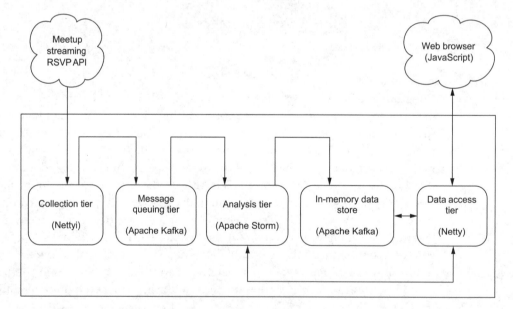

Figure 9.1   Architectural blueprint with technology choices

For each technology I will guide you through what we need to do and explain how it fits in, my reasoning for choosing it, and what to consider as you take this to the next level. Based on what we've learned in previous chapters, for some tiers you already know what the various options are, and in those cases I will refer to our earlier discussions.

We're now ready to begin building our pipeline! This would be a good time to refill your coffee, get comfortable, and (when you're ready) continue on to our discussion of the collection tier.

9.1   *The collection tier*

As we embark on building the collection tier we need to take into consideration how we are going to ingest the data. Ingestion of data comes in many flavors, from the standpoint of the protocols that are used, the data format, and whether the data is being pushed or pulled to or from the collection service. The Meetup Streaming RSVP API is exposed over HTTP using chunking and as a WebSocket service. In both cases, the data is returned as JSON. For our collection service we'll use the WebSockets API and build the client using the Netty (http://netty.io) library.

I chose the WebSockets API and Netty for two reasons: WebSockets is an efficient protocol, as we've discussed in earlier chapters; and we'll be using both WebSockets and Netty latter in this chapter when we build the Data Access API. This allows us to keep things simpler—which is something you should always aim to achieve when you build a real streaming pipeline. Why Netty and not technology X? Netty offers fantastic performance and provides low-level control over the interaction with a client. Both features are important and are discussed in more detail in the Data Access API section later in the chapter.

9.1.1   *Collection service data flow*

When building our collection service, we want to take into consideration the following capabilities:

- Managing the connection to the Meetup API
- Ensuring that we don't lose data
- Integrating with the message queuing tier

The overall flow of how this will be put together is illustrated in figure 9.2.

As you will see, our collection service is composed of a handful of classes. It integrates with Apache Kafka, which we will use in the message queuing tier. To ensure that we don't lose data we're also going to implement the HML that we discussed in chapter 2.

Let's take a deeper look at the key methods illustrated in figure 9.2, starting with the `initialize` method being called on the `HybridMessageLogger`—it's a pretty straightforward method, as shown listing 9.1.

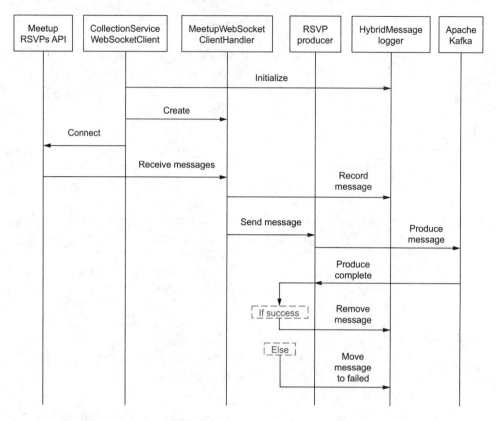

Figure 9.2   High-level sequence diagram of collection service

---

See
chapter 2
section
2.3.3

Listing 9.1   Initializing `HybridMessageLogger`

```
public class CollectionServiceWebSocketClient {
    public static void main(String[] args) throws Exception {
        try{
            HybridMessageLogger.initialize();
        }catch (Exception exception){
            System.err.println("Could not initialize HybridMessageLogger!");
            exception.printStackTrace();
            System.exit(-1);
        }
    }
}

final class HybridMessageLogger {

    private static RocksDB transientStateDB = null;
    private static RocksDB failedStateDB = null;
    private static Options options = null;
    private static Path transientPath = new File("state/transient").toPath();
    private static Path failedPath = new File("state/failed").toPath();
```

**Initialize the Hybrid-
MessageLogger. If it
fails, exit.**

```
static void initialize() throws Exception {        ←  Make sure RocksDB
    RocksDB.loadLibrary();                              is loaded.
    options = new Options().setCreateIfMissing(true);

    try {                                          ┐  Ensure the directories to
        ensureDirectories();                       ←  store the data are created
        transientStateDB = RocksDB.open(options,   ┘  and create the databases.
        transientPath.toString());
        failedStateDB = RocksDB.open(options, failedPath.toString());
    } catch (RocksDBException | IOException e) {
        e.printStackTrace();
        throw new Exception(e);
    }
}
                                                      ┐  Create the
                                                      │  directories on
private static void ensureDirectories() throws IOException {  ←  the filesystem if
                                                      │  they don't exist.
    if(Files.notExists(transientPath)){           ┘
        Files.createDirectories(transientPath);
    }
    if(Files.notExists(failedPath)){
        Files.createDirectories(failedPath);
    }
}
}
```

When the CollectionServiceWebSocketClient class calls the initialize method, the HybridMessageLogger does several important things. First and foremost, it loads the RocksDB library. If you're not familiar with it, RocksDB (http://rocksdb.org) is a fast, embeddable key-value store, perfect for our HML feature. You may want to consider other options, but keep in mind the following minimum requirements for the backing store:

- *It must be fast*—There are going to be a lot of writes and reads; every message received from the Meetup API will result in at least two reads and two writes.
- *The data must be stored locally*—If we must send the data over the network, we may lose it and have to account for that.

After loading RocksDB, we ensure that the directories it's going to store data in are created. Finally, we create two databases, transientStateDB and failedStateDB. transientStateDB will hold the messages that we receive until we get an acknowledgment from Kafka that it was stored successfully. In the case where it fails being sent to Kafka, the message will get moved to the failedStateDB. We'll look at how this works soon.

We're ready to create our MeetupWebSocketClientHandler and connect to the Meetup Streaming RSVP API. As shown in the following listing, for the most part this is boilerplate Netty code.

See
chapter 2
section
2.1.5

Listing 9.2   Creating WebSocket handler and connecting to Meetup API

```java
public class CollectionServiceWebSocketClient {
    public static void main(String[] args) throws Exception {    ❶ Set up the URL
                                                                   we're going to
        ... initializing HybridMessageLogger .......              connect to.

        final String URL =  0 < args.length? args[0] :
    "ws://stream.meetup.com/2/rsvps";
        URI uri = new URI(URL);
        final int port = 80;                                     ❷ Create the
                                                                   handler that
        try {                                                      will receive
            final MeetupWebSocketClientHandler handler =           all messages.
                new MeetupWebSocketClientHandler(
                    WebSocketClientHandshakerFactory.newHandshaker(
                        uri, WebSocketVersion.V13, null, false,
    new DefaultHttpHeaders()));

            Bootstrap b = new Bootstrap();
            b.group(group)
                .channel(NioSocketChannel.class)
                .handler(new ChannelInitializer<SocketChannel>() {
                    @Override
                    protected void initChannel(SocketChannel ch) {
                        ChannelPipeline p = ch.pipeline();       ┌ Add our
                        p.addLast(                               │ handler to
                            new HttpClientCodec(),               │ the Netty
                            new HttpObjectAggregator(8192),      │ pipeline.
    WebSocketClientCompressionHandler.INSTANCE,
                            handler);
                    }
                });
            Channel ch = b.connect(uri.getHost(),                ┌ Connect to the
                    port).sync().channel();                      │ Meetup API and
            handler.handshakeFuture().sync();                    │ wait till the
                                                                 │ handler exists.
            .... handle WebSocket ping and bye messages .....

        } finally {
            group.shutdownGracefully();
        }
    }
}
```

After we initialize the HybridMessageLogger we're ready to set up the WebSocket handler. The way Netty works is we create a handler, which is responsible for all the interaction with its network peer, and then add it to a pipeline. But first we need to set up the URL we will connect to. To aid in debugging or switching the WebSocket server we connect to, ❶ we check to see if there are any command-line arguments passed in. If there are, we use the first one as the URL to connect to. Note that we're not checking to see if it's a valid URL—you would want to do that in a real setting. If no arguments are passed, then we default to using the current Meetup Streaming RSVP API. We're then ready to set up the `MeetupWebSocketClientHandler` ❷, passing into it a `WebSocketClientHand-shaker`, a standard Netty object that takes care of the lower-level protocol interactions so we can concentrate on the message handling. When the `MeetupWebSocketClient-Handler` is constructed, it in turn constructs an `RSVPProducer`.

Listing 9.3 Construction of `MeetupWebSocketClientHandler` and `RSVPProducer`

```
class MeetupWebSocketClientHandler extends SimpleChannelInboundHandler<Object>
{
    private final static RSVPProducer rsvpProducer = new RSVPProducer();    <─┐
}
```
Constructing the RSVPProducer

Constructing the `RSVPProducer` will in turn construct a `KafkaProducer`, as shown next.

Listing 9.4 Construction of `RSVPProducer`

```
final class RSVPProducer {
    private static KafkaProducer<byte[], byte[]> kafkaProducer;      Set up the producer
                                                                     properties that are
    RSVPProducer() {                                                 required
        Properties producerProperties = new Properties();    <─┘
        producerProperties.put(ProducerConfig.BOOTSTRAP_SERVERS_CONFIG,
            "localhost:9092");

        producerProperties.put(ProducerConfig.CLIENT_ID_CONFIG,
            "meetup-collection-service-kafka");

        producerProperties.put(ProducerConfig.KEY_SERIALIZER_CLASS_CONFIG,
            "org.apache.kafka.common.serialization.ByteArraySerializer");

        producerProperties.put(ProducerConfig.VALUE_SERIALIZER_CLASS_CONFIG,
            "org.apache.kafka.common.serialization.ByteArraySerializer");

        kafkaProducer = new KafkaProducer<>(producerProperties);    <─┐
    }
}
```
Construct the KafkaProducer

Now that we've constructed `MeetupWebSocketClientHandler`, `RSVPProducer`, and `KafkaProducer`, we can add `MeetupWebSocketClientHandler` to the Netty pipeline.

Listing 9.5 Adding the handler to the Netty pipeline

```
public class CollectionServiceWebSocketClient {
  public static void main(String[] args) throws Exception {

        ... Initializing HybridMessageLogger .......
        ... Setup the URL ...
            ... Construct the Handler ...

        Bootstrap bootStrap = ......
        bootStrap.channel(NioSocketChannel.class)
            .handler(new ChannelInitializer<SocketChannel>() {
```

Build up
the channel
pipeline ❶

```
                    @Override
                    protected void initChannel(SocketChannel ch) {
                      ChannelPipeline channelPipeline = ch.pipeline();
                      channelPipeline.addLast(new HttpClientCodec());
                         channelPipeline.addLast(new HttpObjectAggregator(8192));
                         channelPipeline.addLast(
                                 WebSocketClientCompressionHandler.INSTANCE);
                      channelPipeline.addLast(handler);
                    }
                  });                                           ❷  Set the handler
      }
    }
```

Constructing the channel pipeline ❶ and setting the handler ❷ ensure that it is the
last handler when messages are being processed off the wire by Netty. At this point, all
the required objects are configured and the pipeline is ready to go—the only thing
left to do is to connect to the Meetup Streaming API, as shown next.

Listing 9.6   Connecting to Meetup API

```
public class CollectionServiceWebSocketClient {
    public static void main(String[] args) throws Exception {

        ... Initializing HybridMessageLogger .......
        ... Setup the URL ...                              After this line executes the
        ... Construct the Handler                          collection service will be
        ... Setup the pipe line                            running in the background,
                                                           connected to the Meetup API
        Channel ch = b.connect(uri.getHost(),              and ready to receive messages.
            port).sync().channel();

        handler.handshakeFuture().sync();                  This line will block
                                                           waiting for the
        .... handle WebSocket ping and bye messages .....  collection service
                                                           to shut down.
    }
}
```

MeetupWebSocketClientHandler will receive a new message for every new RSVP that
Meetup.com receives. The following listing illustrates how we will handle these new
messages in the channelRead0 function of MeetupWebSocketClientHandler.

Listing 9.7   Message handling in MeetupWebSocketClientHandler

Declaration
of the
channelRead0
method ❷

```
class MeetupWebSocketClientHandler extends              ❶  Set up the
    SimpleChannelInboundHandler<Object>                     members of
{                                                           this class we
    private final WebSocketClientHandshaker handshaker;     will be using
    private ChannelPromise handshakeFuture;
    private final static RSVPProducer rsvpProducer = new RSVPProducer();

    @Override
    public void channelRead0(ChannelHandlerContext ctx, Object msg)
                            throws Exception {
        Channel channel = ctx.channel();
```

Check to verify that WebSocket handshaking is complete ❸

```
if (!handshaker.isHandshakeComplete()) {
    //we are now connected.
    handshaker.finishHandshake(channel, (FullHttpResponse) msg);
    handshakeFuture.setSuccess();
    return;
}
```

Sanity check that the message we receive is not an erroneous HttpMessage ❹

```
if (msg instanceof FullHttpResponse) {
    //if we get this, it is an error and we should throw an
    //exception

    FullHttpResponse response = (FullHttpResponse) msg;
    throw new IllegalStateException(
        "Unexpected FullHttpResponse (getStatus=" + response.status()
        + ", content=" +
        response.content().toString(CharsetUtil.UTF_8) + ')');
}
```

Use a random value for the message key ❺

```
WebSocketFrame frame = (WebSocketFrame) msg;
if (frame instanceof TextWebSocketFrame) {
    TextWebSocketFrame textFrame = (TextWebSocketFrame) frame;

    final String messageKey =
        new UUID(random.nextLong(),
            random.nextLong()).toString();
```

Read all data from the ByteBuf ❻

```
    //this is the message we want,we can take the
    //data and send it to to the next tier.
    //First we need to read the bytes from the ByteBuf
    final byte[] messagePayload =
            new byte[textFrame.content().readableBytes()];
    textFrame.content().readBytes(messagePayload);
```

The receiver side of the HML protocol ❼

```
    HybridMessageLogger.addEvent(messageKey,messagePayload);

    rsvpProducer.sendMessage(messageKey,messagePayload);
} else if (frame instanceof CloseWebSocketFrame) {
    //we are being asked to close.
    channel.close();
    rsvpProducer.close();
}
}
}
```

Send the message to the queuing tier

Close things down if necessary

There are several objects that we need to set up in ❶ for our handler to use—the RSVPProducer we already looked at, and the others we get from the constructor. The real work for this class begins at ❷, the method that will be called by Netty every time a message is received on the wire. However, because we received the message doesn't mean we're ready to process it. First, as in ❸, we need to check and see if the Web-Socket protocol handshake is complete. Once it is, we need to do one final check in ❹ before we can safely process the data. It's an error to receive an HTTP message, which is why we throw an exception if we encounter one. Finally, we get to ❺ where we're ready to do the real work. First, we create the message key that will be used for both our receiver-based logging and for Kafka. We're using a UUID in ❺ because that's guaranteed to be unique, which will allow for the data to be evenly distributed by Kafka. After the key is generated, we're ready to read the data that was passed in the message. ❻ is the safest way to read the data that was received. In this case it will be an RSVP, represented as JSON. With the key and the message in hand we can now call the HybridMessageLogger in ❼.

That addEvent call on HybridMessageLogger is straightforward, as shown here.

See
chapter 2
section
2.3.3

Listing 9.8 HybridMessageLogger addEvent call

```
static void addEvent(final String eventKey, final byte[] eventData)throws
    Exception{
  try {
    final byte[] keyBytes = eventKey.getBytes(StandardCharsets.UTF_8);

    byte[] value = transientStateDB.get(keyBytes);

    if (value == null) {
      transientStateDB.put(keyBytes, eventData);
    }
  } catch (RocksDBException ex) {
    //would want to log this occuring
    throw new Exception(ex.getMessage());
  }
}
```

Get the bytes for the key using UTF-8 encoding ❶

See if the key already exists in the DB ❷

If the key doesn't exist, add it ❸

We get a key as a byte array ❶ and we check to see if there is already a value associated with this key in the DB ❷. We do this so that we can be sure we only store one value associated with a key at a time. If we don't find the value, then in ❸ we add a key-value pair to the DB. Remember, we need to make sure that code is called *before* we send the data to the message queuing tier. This is the RBML part of the HML algorithm discussed in chapter 2. We're still at risk for losing data if something happens to our collection service between ❺ and ❼ in listing 9.7. There are ways to solve this, but that's outside of the scope of this chapter; I'll leave it as an exercise for you to undertake. Now that the data is safely stored in ❽ of listing 9.7, we can finally pass the message to the rsvpProducer so that it can be sent to Kafka.

Listing 9.9 RSVPProducer send method

```
final class RSVPProducer {
  private static KafkaProducer<byte[], byte[]> kafkaProducer;
  private static final String messageTopic = "meetup-raw-rsvps";

  ... Constructor ...
  void sendMessage(final String messageKey, final byte[] message) {

    ProducerRecord<byte[],byte[]> producerRecord =
      new ProducerRecord<>(messageTopic,
      messageKey.getBytes(StandardCharsets.UTF_8),
      message);
```

The topic name for Kafka

Create a ProducerRecord with topic, key, and value ❶

```
kafkaProducer.send(producerRecord,
    new TopicCallbackHandler(messageKey));
  }
}
```

Send the data ❷ to Kafka

We create topic name that we're going to send data to in Kafka. We'll want to create this topic in the next section when we set up Kafka. If you run the example code, this topic will need to be created beforehand. Before we can send data to Kafka, we need to create a ProducerRecord, as is done in ❶. With that created, we can now send the data to Kafka, as in ❷. Note that we're passing a callback to Kafka. This callback is passed the eventKey when constructed and will allow us to determine if the data was sent and received successfully. There's not that much to this callback, as you can see in the next listing.

Listing 9.10 `TopicCallbackHandler`

```
private final class TopicCallbackHandler implements Callback {
  final String eventKey;

  TopicCallbackHandler(final String eventKey){
    this.eventKey = eventKey;
  }

  @Override
  public void onCompletion(RecordMetadata metadata, Exception exception) {
    if (null == metadata){
      //mark record as failed
      HybridMessageLogger.moveToFailed(eventKey);
      //log the exception or do something with it.
    }else{
      //remove the data from the localstate
      try {
          HybridMessageLogger.removeEvent(eventKey);
      } catch (Exception e) {
        //this should be logged...
      }
    }
  }
}
```

The constructor taking the eventKey ❶ as an argument

If the metadata is null, there ❷ was an error

Data sent successfully, ❸ remove from DB

In ❶ the eventKey is passed into the constructor so that it can be used later in onCompletion to remove the data from the message logging. If there is a problem sending the data to Kafka or in Kafka handling it, then RecordMetadata will be null ❷, and we'll need to handle that. If the data is successfully stored in Kafka, then we remove it ❸ from HybridMessageLogger—this is the sender-based message logging algorithm from chapter 2 that is the other half of HML.

I know this may have seemed like a lot, but overall there's not that much code. With this code in hand we now have a fully functional collection service that can connect to the Meetup Streaming RSVP API, ensure no data loss, and send the data to the message queuing tier. Before we can run this service, we need to go through the next section on Apache Kafka.

This is a good time to take a break; when you come back we'll continue with the message queuing tier.

9.2   Message queuing tier

See chapter 3 section 3.2

In chapter 3 I discussed options for the message queuing tier. For our streaming Meetup pipeline, I've chosen to use Apache Kafka. It's scalable, has great performance, provides data durability guaranteed to ensure that we don't lose data, and is easy to integrate with. In this section we're not going to go into detail covering all the aspects of Kafka. We'll cover enough to help us work through the example we're building in this chapter. For more information, consult the project homepage at http://kafka.apache.org or books, such as Neha Narkhede's *Kafka: The Definitive Guide* (O'Reilly, 2016).

9.2.1   Installing and configuring Kafka

We need to download and install Kafka from http://kafka.apache.org/downloads.html. For this chapter we'll use version 0.10.0.1, though a later version should work. Let's start by downloading and installing it by executing the following listing's commands in a terminal window.

Listing 9.11   Downloading and installing Apache Kafka

```
$> wget http://www-us.apache.org/dist/kafka/0.10.0.1/kafka_2.11-0.10.0.1.tgz
$> tar -xvf kafka_2.11-0.10.0.1.tgz
```

There are four main concepts to keep in mind when working with Kafka:

- Producers
- Topics
- Brokers
- Consumers

Producers, brokers, and consumers are much like what was covered in chapter 3. Topics in Kafka are the logical grouping that messages are written to and read from. A producer sends messages to a broker for a particular topic, and the consumer requests to read data from a particular topic. In ❶ in listing 9.9, you can see that we're going to be producing data to the `meetup-raw-events` topic. Therefore, the next thing we need to do is start Kafka and create the topic by executing the commands in the following listing.

Listing 9.12   Starting Kafka, Apache ZooKeeper, and creating a topic

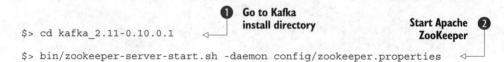

❶ Go to Kafka
install directory

Start Apache ❷
ZooKeeper

```
$> cd kafka_2.11-0.10.0.1

$> bin/zookeeper-server-start.sh -daemon config/zookeeper.properties
```

Start   ⬧→ `$> bin/kafka-server-start.sh -daemon config/server.properties`
Kafka ➌
              `$> bin/kafka-topics.sh --zookeeper localhost:2181 --create \`   ←┐ **Create**
              `    --topic meetup-raw-rsvps --partitions 1 --replication-factor 1` ➍ **a topic**

              `$> Created topic "meetup-raw-rsvps".` ←┐ **The expected**
                                                     ➎ **output**

The first thing we need to do on the command line is change to the Kafka install directory ➊. Kafka relies on Apache ZooKeeper for storing some metadata about topics and brokers; so we start ZooKeeper ➋; next we start up a single Kafka broker ➌. At this point both ZooKeeper and Kafka are running, and we can create ➍ the topic we need. ➎ is the output we should expect to see, indicating that the topic was created. To verify the topic was created we can use the command in the following listing.

Listing 9.13 Kafka list topics command

```
$> bin/kafka-topics.sh --zookeeper localhost:2181 --list    ←┐ The list topics
                                                              │ command
    meetup-raw-rsvps            ←┐ The expected
$>                               │ output
```

Kafka is now completely ready for our collection service to start sending messages to it.

9.2.2 *Integrating the collection service and Kafka*

With the collection service done and Kafka up and running, we can integrate the two to make sure things are working up to this point. To test the integration, we need to open two console windows. In the first one we'll run the Kafka console consumer. To do so, enter the command in the following listing.

Listing 9.14 Running the Kafka console consumer

```
$> bin/kafka-console-consumer.sh --zookeeper localhost:2181\
 --topic meetup-raw-rsvps
```

Because there are no messages currently being written to this topic, nothing will appear to happen after executing the preceding command. That's perfectly normal—it's waiting for messages to arrive before displaying them.

The next step is to get our consumer services running. We can do that with the commands in the following listing.

Listing 9.15 Building and running the collection service

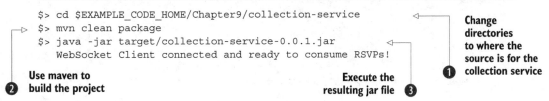

```
$> cd $EXAMPLE_CODE_HOME/Chapter9/collection-service      ←┐  Change
$> mvn clean package                                       │  directories
$> java -jar target/collection-service-0.0.1.jar      ←┐   │  to where the
    WebSocket Client connected and ready to consume RSVPs! │  source is for the
                                                           ➊  collection service
```

Use maven to   **Execute the**
➋ **build the project**   **resulting jar file** ➌

After changing to the directory where you have the source for the collection service ❶, we will build the project using Apache Maven ❷. If you don't have Maven installed, you can download it or find install instructions for your OS at https://maven.apache.org. We're now ready to run the collection service by executing ❸. After you see the message that the collection service is connected and ready, you should start to see messages appear in the Kafka consumer output window. An example of an RSVP message is shown in figure 9.3. Feel free to quit the collection service by pressing Ctrl-C in the console window.

```json
{
    "venue": {
        "venue_name": "mount work regional park, munn road parking lot",
        "lon": -123.464233,
        "lat": 48.500511,
        "venue_id": 12469832
    },
    "visibility": "public",
    "response": "yes",
    "guests": 2,
    "member": {
        "member_id": 215609630,
        "photo": "http://photos3.meetupstatic.com/photos/member/a/6/6/4/
            thumb_261342596.jpeg",
        "member_name": "Amanda Dunn"
    },
    "rsvp_id": 1638108083,
    "mtime": 1478449512424,
    "event": {
        "event_name": "Sunday Funday Mt. Work from Munn Road",
        "event_id": "235366173",
        "time": 1478455200000,
        "event_url": "http://www.meetup.com/Random-Activities-Victoria/events/
            235366173/"
    },
    "group": {
        "group_topics": [{
            "urlkey": "bike",
            "topic_name": "Bicycling"
          },
          {
            "urlkey": "local-activities",
            "topic_name": "Local Activities"
          }
        ],
        "group_city": "Victoria",
        "group_country": "ca",
        "group_id": 20896502,
        "group_name": "Random Activity Tribe",
        "group_lon": -123.29,
        "group_urlname": "Random-Activities-Victoria",
        "group_state": "BC",
        "group_lat": 48.46
    }
}
```

Figure 9.3 Example Meetup RSVP JSON message

9.3 Analysis tier

See
chapter 4
section
4.2

There are numerous options for the analysis tier—in fact, the last check I did revealed more than 25 different products all vying to be the tool of choice for this tier. We'll use Apache Storm; it's mature, easy to install and get running, and provides tuple-at-a-time streaming, which is what we need for Meetup RSVP processing. The architecture we've been working with throughout the book and bringing to life in this chapter enables us to switch out the technology in a given tier, in many cases without impacting any other tier. To develop a deeper understanding of the distributed stream-processing engines, as an exercise after working through this chapter, I encourage you to swap out Storm for a different product. After installing and setting up Storm, we're going to build a topology that computes the top *n* topics for all RSVPs. *Topics* in the Meetup world are keywords that an organizer chooses that help identify their group. The following listing shows a snippet of a JSON RSVP message.

Listing 9.16   RSVP Message JSON

```
{
   ...
   "group": {                              ◁─────┐  Other parts of
     "group_topics": [                         ❶  the message
       ...
       {                                    ◁─────┐ The group topics,
                                                    many eliminated
         "urlkey": "exploring-the-city",      ❷  for brevity
         "topic_name": "Exploring the City"
       },
       {
         "urlkey": "charity-work",
         "topic_name": "Charity Work"
       },
       {
         "urlkey": "local-businesses",
         "topic_name": "Local Businesses"
       },
       {
         "urlkey": "cultural-activities",
         "topic_name": "Cultural Activities"
       }                                               Metadata about
     ],                                                   the group
     ...
     "group_name":"AMS Connected - Where Expats & Amsterdam Truly Unite"  ◁──┘
     ...
   }
}
```

The data in the full JSON object is much richer than the snippet shown in the previous listing. There's more metadata ❶ about the event and the venue—perhaps there are other JSON elements that you may want to do further analysis on. For example, there is geo information; with that you can drop pins on a map and show which parts of the world have the most activity. Or perhaps you can perform an analysis that shows the trending topics by geography. ❷ shows some of the group topics; currently Meetup allows group organizers to choose 15 topics to help describe and categorize their group. For the analysis in this tier we'll use all the group topics we receive in the JSON message.

9.3.1   Installing Storm and preparing Kafka

For this chapter we'll only be installing Storm for development and running locally. To find out more about installing Storm for production usage, see Sean T. Allen's *Storm Applied* (Manning, 2015). To perform our setup, enter these commands.

Listing 9.17   Installing Apache Storm

```
Download Storm from: http://storm.apache.org/downloads.html    ◁┐  Download from
Decompress the downloaded distribution                            │  this site and
                                                                  │  decompress
Or use the following commands:

$> wget http://www-us.apache.org/dist/storm/apache-storm-1.0.2/apache-storm-
    1.0.2.tar.gz                                  ◁┐  Command-line way
                                                   │  to download and
$> tar -xvf apache-storm-1.0.2.tar.gz              │  install
```

With Storm downloaded and uncompressed, we're ready to create the Kafka topic we will use for storing the results of our analytics. The steps required to set up Kafka are given next.

Listing 9.18   Setting up Kafka for Analysis topic

```
$> cd $KAFKA_INSTALL_DIR                                                    ◁┐
$> bin/kafka-topics.sh --zookeeper localhost:2181 --create --topic meetup-  │
    topn-rsvps --partitions 1 --replication-factor 1    ◁┐                   │
$> Created topic "meetup-topn-rsvps"                     │   Navigate to the │
                                                         │   install directory│
                     Create the topic where the         │       for Kafka   │
                     analysis is going to be written     │
```

First, navigate to the directory where you have installed Kafka and then create a new topic, called "meetup-topn-rsvps." That's it—Storm is ready to use, and we've created our topic in Kafka. We're ready to go through the code we will use to do the analysis and then see how to integrate it with the rest of the pipeline we've built so far.

9.3.2   Building the top n Storm topology

Let's get down basic Storm terminology first. In Storm, a *topology* is a directed acyclic graph (DAG) that's composed of spouts and bolts. *Spouts* ingest data from a data source; think of them as pouring data into the topology; each element of the data is then represented as a tuple in the topology. *Bolts* are the pieces of business logic that perform a computation on the data—perhaps an aggregation, filtering, augmentation, or even writing the data to a different destination. To learn more about Storm, I again recommend Sean T. Allen's *Storm Applied* (Manning, 2015).

To perform the top *n* analysis of the Meetup RSVPs we have several options when it comes to building a Storm topology. We could opt for the traditional multi-bolt approach illustrated in figure 9.4.

See chapter 5 section 5.3.3

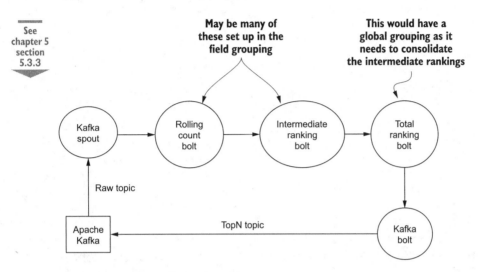

Figure 9.4   Multi-bolt approach to top *n*

With the approach in figure 9.4 we end up being able to have many rolling count and intermediate ranking bolts, each producing a fragment of the final result. You can have many of these bolts, and when using the field grouping, the data will get shuffled around the cluster. This changes for the *total ranking bolt* because its job is to produce the complete ranking for all the intermediates. Therefore there can only be one instance of it, and the grouping is global—so all the data from all the intermediates will go to the total ranking bolt so can it can produce the top *n* ranking across all the data. To see this type of topology fleshed out, the code for the Apache Storm Starter project, available at https://github.com/apache/storm/tree/master/examples/storm-starter, has a good example of this.

Instead of using that type of topology, I've opted to use a simpler approach and build a topology using the stream summarization functionality of stream-lib, found at https://github.com/addthis/stream-lib. Our topology, illustrated in figure 9.5, looks a little different.

By using this architecture we eliminate several bolts, and you get a chance to see how to integrate stream-lib, which is a library rich in functionality used to perform analysis of a stream of data. Our topology has the same single bolt that all data travels through to perform the computation, the same as the multi-bolt architecture. The difference here is that we are using the stream-lib library. In this library we will be using

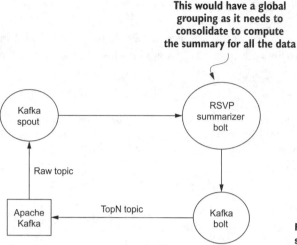

This would have a global grouping as it needs to consolidate to compute the summary for all the data

Figure 9.5   Topology using streaming summary

the StreamSummary class. This class provides a top *n* algorithm based on the space-saving algorithm and the stream-summary data structure described in "Efficient Computation of Frequent and Top-k Elements in Data Streams" by Metwally, Agrawal, and Abbadi (http://dl.acm.org/citation.cfm?id=2131596). This library also contains other algorithms solely designed for operating on data in a stream. I encourage you (as an exercise) to replace StreamSummary with the Count-Min-Sketch algorithm that stream-lib also includes (CountMinSketch) to see first-hand how the algorithm we discussed in chapter 5 works with live data. Here is the basic setup of our topology.

Listing 9.19   Setup of topology

```
private StormTopology buildTopolgy() {

  final TopologyBuilder tb = new TopologyBuilder();

  tb.setSpout("kafka_spout", new KafkaSpout<>(getKafkaSpoutConfig()), 1);

  tb.setBolt("rsvpSummarizer", new MeetupRSVPSummaryBolt(),
      1).globalGrouping("kafka_spout", STREAM_NAME);

  tb.setBolt("summarySerializer", new
      KafkaBolt(),1).shuffleGrouping("rsvpSummarizer");

  return topologyBuilder.createTopology();
}
```

Create the TopologyBuilder ❶

Add the spout to the topology ❷

Add the SummaryBolt to the topology ❸

Add the KafkaBolt to the topology ❹

Return the defined topology ❺

First, ❶ we create the ToplogyBuilder we're going to use to add the spout and bolts to. Next, ❷ we set up the spout—the built-in KafkaSpout—we're going to use. (Shortly we'll go over the getKafkaSpoutConfig code). The next step ❸ is setting up our bolt that does the top *n* computation using stream-lib. In ❹ we add a KafkaBolt to the topology. As a last step, ❺ we return the bolt. As we will see, returning the bolt allows us to wrap this method in a call to either run the topology in local mode or submit it to a cluster. The getKafkaSpoutConfig method wraps up the setup code.

Listing 9.20 getKafkaSpoutConfig

```
private KafkaSpoutConfig<String, String> getKafkaSpoutConfig() {
    return new KafkaSpoutConfig.Builder<>(
        getKafkaConsumerProps(),
        getKafkaSpoutStreams(),
        getTuplesBuilder(),
        getRetryService())
    .build();
}
```

Construct the KafkaSpout-Config builder

Generate the consumer properties

Generate the streams for the spout

Generate the tuple builder

Build the SpoutConfig

Generate the retry service

KafkaSpoutConfig builder takes several parameters, and for each we call a separate method to construct it. The getKafkaConsumerProps is shown next.

Listing 9.21 getKafkaConsumerProps

```
private Map<String, Object> getKafkaConsumerProps() {
    Map<String, Object> props = new HashMap<>();
    props.put(KafkaSpoutConfig.Consumer.ENABLE_AUTO_COMMIT, "true");
    props.put(KafkaSpoutConfig.Consumer.BOOTSTRAP_SERVERS, "127.0.0.1:9092");
    props.put(KafkaSpoutConfig.Consumer.GROUP_ID, TOPIC_NAME + "-group");

    props.put(KafkaSpoutConfig.Consumer.KEY_DESERIALIZER,
        "org.apache.kafka.common.serialization.StringDeserializer");

    props.put(KafkaSpoutConfig.Consumer.VALUE_DESERIALIZER,
        "org.apache.kafka.common.serialization.StringDeserializer");

    return props;
}
```

❶ Create the map that holds all the properties

❷ Set the key deserializer

❸ Set the value deserializer

The map ❶ is used to hold all the properties. The first property ENABLE_AUTO_COMMIT tells the Kafka Consumer to commit offsets because we're processing messages automatically. The second property is setting the Kafka servers—in this case, we have it running on our localhost. The third property sets the GROUP_ID, which is used by the Kafka Consumer to uniquely identify the consumers in a group. The other interesting properties that we're setting are ❷ and ❸. Here we set up the Deserializers. This will let the Kafka Consumer know that we want to take the byte arrays that the key and value will be in and deserialize them into strings. Continuing to dive into the Kafka-SpoutConfig Builder parameters, we look at getKafkaSpoutStreams next.

Listing 9.22   `getKafkaSpoutStreams`

```
private KafkaSpoutStreams getKafkaSpoutStreams() {
    final Fields outputFields = new Fields( "topic",
                                            "partition",
                                            "offset",
                                            "key",
                                            "value");

    return new KafkaSpoutStreamsNamedTopics.Builder(
                outputFields,
                STREAM_NAME,
                new String[]{TOPIC_NAME})
            .build();
}
```

① Declare the output fields the spout will emit

② Create the stream with the fields

Each field ① that the spout will be emitting will be available in the tuple that a bolt receives. The tuple will automatically be created with these fields, and each one will contain the associated value. We tie together ② the output fields with the stream name and return a KafkaSpoutStream. The getTuplesBuilder shown in the following listing is the next argument to the KafkaSpoutConfig builder.

Listing 9.23   `getTuplesBuilder`

```
private KafkaSpoutTuplesBuilder<String, String> getTuplesBuilder() {

    return new KafkaSpoutTuplesBuilderNamedTopics.Builder<>(
                new TopicTupleBuilder(TOPIC_NAME)
    ).build();
}
```

Constructs the tuple builder

Now let's look at the code for the TopicTupleBuilder.

Listing 9.24   `TopicTupleBuilder`

```
public class TopicTupleBuilder<K,V> extends KafkaSpoutTupleBuilder<K,V> {

    public TopicTupleBuilder(String... topics) {
        super(topics);
    }

    @Override
    public List<Object> buildTuple(ConsumerRecord<K, V> consumerRecord) {
        return new Values(consumerRecord.topic(),
                consumerRecord.partition(),
                consumerRecord.offset(),
                consumerRecord.key(),
                consumerRecord.value());
    }
}
```

Takes any number of topics and calls the base class

① Determines how to construct a tuple from the Kafka ConsumerRecord

① is where the meat of this builder is, and that's a pretty simple method of creating a tuple from the Kafka ConsumerRecord. You could do something more complex here if you needed to populate the tuple with more information than is provided from the `ConsumerRecord`—perhaps look up some other metadata and add it to the tuple. We're almost done with the arguments to the `KafkaSpoutConfig`. The last one is `get-RetryService`.

Listing 9.25 `getRetryService`

```
private KafkaSpoutRetryService getRetryService() {
    return new KafkaSpoutRetryExponentialBackoff(         ◁─┐  Creates and sets up the
        TimeInterval.microSeconds(500),                       ExponentialBackoff
        TimeInterval.milliSeconds(2),                         service
        Integer.MAX_VALUE,
        TimeInterval.seconds(10));
}
```

This method creates and configures the built-in `ExponentialBackoff` object. For more information on what that retry implementation does, look at the Storm API document for that class at http://mng.bz/hdrh.

That covers ❷ of listing 9.18. Now let's look at `MeetupRSVPSummaryBolt`, the class that ❸ in listing 9.18 is creating. That class is where the magic is occurring to compute the top *n*, the body of which is as follows.

Listing 9.26 `MeetupRSVPSummaryBolt`

```
public class MeetupRSVPSummaryBolt extends BaseRichBolt {          ❶ Construct the
    private static StreamSummary<String> streamSummary =    ◁─┘      StreamSummary
        new StreamSummary<>(100000);                                 object
    private static final ObjectMapper objectMapper = new ObjectMapper();

    @Override
    public void execute(Tuple tuple) {                       ❸ Get the RSVP
        String jsonString = tuple.getString(4);       ◁──       JSON object

        JsonNode root = objectMapper.readTree(jsonString);
        JsonNode groupTopics = root.get("group").get("group_topics");

        Iterator<JsonNode> groupTopicsItr = groupTopics.iterator();

        while(groupTopicsItr.hasNext()){
            JsonNode groupTopic = groupTopicsItr.next();
            final String topicName = groupTopic.get("topic_name").asText();
            streamSummary.offer(topicName);                   ◁─┐ Add it to the
        }                                                   ❼    summary

        collector.emit(new Values(UUID.randomUUID().toString(),   ◁─┐ Get the current
            objectMapper.writeValueAsString(                    ❽    top 10
                    streamSummary.topK(10)
                    )
        ));

    }
}
```

Construct the JSON serializer ❷

Convert it to a JsonNode ❹

Get the group_topics node ❺

Get the group topic name ❻

❶ and ❷ are setting up the objects we need to use for the stream summary and manipulation of JSON data respectively. ❸ gets the JSON RSVP data from the tuple. In listing 9.21 ❶, we declared that the RSVP data was the value from Kafka and would be in position number 4 of the tuple. After we have the JSON representation of the data, we can get the group_topics node, iterate over it, in ❻ get the topic name, and in ❼ add it to the running summary. After going over all the topics, we're ready to emit data. ❽ does that, emitting a UUID and a JSON representation of the running top 10 results from the stream summary. A snippet of the JSON from the summary is next.

> **Listing 9.27   Snippet of `StreamSummary` JSON**

```
[
  {
    "item": "Social",
    "count": 59,
    "error": 0
  },
  {
    "item": "Social Networking",
    "count": 45,
    "error": 0
  },
... (8 more items) ...
]
```

This is pretty simple; it has the group topic, represented by the item and the count of times it has occurred and any error counts. We will be revisiting this structure when we build the UI.

The last part of configuring our topology uses a Kafka bolt to do the writing to Kafka. There is a built-in Kafka bolt with Storm, and one is also included in the code sample for this tier. Because we walked through the Kafka-producing code when we talked about the collection service, we'll skip the code listing because, as you can imagine, it looks similar. It's in the code download, so do peruse it if you want to see what it does.

With our topology constructed, we're ready to finally submit it to a Storm cluster. There are two ways to do this. The following listing shows the way we do it for a local and remote cluster.

> **Listing 9.28   Submitting topology to Storm**

```
private void runLocally() throws InterruptedException {          ❶ Create a local
                                                                     cluster
  LocalCluster cluster = new LocalCluster();
  cluster.submitTopology(this.TOPOLOGY_NAME, getConfig(),buildTopolgy());
  stopWaitingForInput();
}

private void runRemotely() throws AlreadyAliveException,
                                  InvalidTopologyException,
                                  AuthorizationException {
```

```
StormSubmitter.submitTopology(TOPOLOGY_NAME,
                              getConfig(),
                              buildTopolgy());
}
```

❷ **Submit to remote cluster**

A local cluster ❶ is perfect for testing and development. All that is involved is creating the cluster and submitting it. `stopWaitingForInput` waits for the user to enter a key on the keyboard to stop everything. ❷ submits the same job to a remote cluster. This cluster can be running on your same machine, but it's a real running Storm cluster. In this exercise we'll only be using the `LocalCluster`, although I encourage you to build a Storm cluster and deploy our topology to it. With all this code in place, we're ready to integrate our analysis tier with the tiers we've already built.

9.3.3 Integrating analysis

At the end of this section we will have the collection service feeding data into Kafka, and the analysis tier reading that raw data, performing a top *n* on it, and writing the results back to Kafka. To get started we will build the analysis tier, found in this chapter's code. To start, execute the commands in the following listing.

Listing 9.29 Building and running the analysis tier

```
$> cd $EXAMPLE_CODE_HOME/Chapter9/analysis-tier
$> mvn clean package
$> $STORM_INSTALL_DIR/bin/storm jar target/analytics-0.0.1.jar
com.streamingdata.analysis.TopMeetupTopicsTopology
```

Change directories to where the source is for the analysis tier

Use Maven to build the project

Use Storm to deploy

Build and submit the topology to a local cluster—remember this is for development and testing. Now that the topology is running, if you don't have the collection service running, go ahead and start it. After a little bit you should see output in the console window you are running the topology from about the tuples being processed. With the collection service, Kafka, and the analysis tier running, let's open one more command window and watch the data that is being emitted to Kafka from the analysis tier. To do that, we'll use the Kafka console consumer—you can start it by using the command in the following listing.

Listing 9.30 Consuming the top *n* topic

```
$> $KAFKA_HOME/bin/kafka-console-consumer.sh --zookeeper localhost:2181
--topic meetup-topn-rsvps
```

Once the console consumer is running, you should start seeing messages emitted. Congratulations! At this point we have the collection service, Kafka, ZooKeeper, and the analysis tier all running. We are consuming a stream of data from Meetup, analyzing it, and making it ready to be consumed. We've brought to life the first five chapters of this book! In the next section we'll walk through the construction of the data access API—one step closer to having a complete pipeline.

9.4   *In-memory data store*

See chapter 6 section 6.2

For this tier we'll use Apache Kafka, which may not be the first technology you think of using for an in-memory data store. I chose it for three reasons:

- We're already using it for our queuing tier, so our architecture stays clean and simple.
- Although Kafka does write data to disk, the speed at which we can read and write is "fast" enough to satisfy our use case for the application we're building.
- The consumer access pattern is conceptually a stream. This allows us to think about accessing the data in the data access tier as a stream, reducing complexity in building our pipeline.

The code listing for this tier is this single line `tb.setBolt("summarySerializer", new KafkaBolt(),1).shuffleGrouping("rsvpSummarizer")` from listing 9.18 **4**. This line wires together the built-in `KafkaBolt` Storm bolt with the output from our `"rsvp-Summarizer"`. As the code in listing 9.25 **8** emits the top 10 JSON data, the `KafkaBolt` receives it and writes it to Kafka. In essence, we get the writing of data to our in-memory store with a single line of code—not a bad deal at all. Not that the number of lines of code should be the deciding factor, but it should count for something. Once you start the analysis tier as described at the end of section 9.3, you will be ready to move on to section 9.5 where you make use of the data.

9.5   *Data access tier*

This is the last step in bringing everything to life. We have everything ready. We're ingesting data, moving it across the queuing tier, analyzing it, and making it ready to be consumed. This is where we make it available for a UI. This section discusses the implementation of the data access API and the basic UI that is built to consume the data from the API. At the end of this section you will have everything running and will be able to visualize a complete end-to-end streaming data pipeline.

For this tier I've implemented the API using Netty as the underlying framework. There are two primary reasons for using Netty. First, it's a high-performance network framework with robust WebSockets support. Secondly, Netty provides low-level access to the underlying connection. If you are to build this for a production setting, then you will need low-level access to be able to understand at a minimum whether the client is still connected and whether the socket is writable. To learn more about Netty, and in particular if you're going to build this for a production environment, I strongly recommend Norman Maurer and Allen Wolfthal's book *Netty In Action* (Manning, 2015).

All right, let's move on to the code. The whole thing starts with basic Netty code to get it running, shown here.

See chapter 7 section 7.2.4

Listing 9.31 StreamingDataService.java

```java
public final class StreamingDataService {
    private static final int PORT = 8080;
    private static final String bindAddress = "127.0.0.1";

    public static void main(String[] args) throws Exception {

        EventLoopGroup bossGroup = new NioEventLoopGroup(1);
        EventLoopGroup workerGroup = new NioEventLoopGroup();

        try {

            ServerBootstrap bootstrap = new ServerBootstrap();
            bootstrap.group(bossGroup, workerGroup)
                .channel(NioServerSocketChannel.class)
                .handler(new LoggingHandler(LogLevel.DEBUG))
                .childHandler(new ChannelInitializer<SocketChannel>() {
            @Override
            public void initChannel(SocketChannel ch){
            ChannelPipeline pipeline = ch.pipeline();
            pipeline.addLast(new HttpServerCodec());
            pipeline.addLast(new HttpObjectAggregator(65536));
            pipeline.addLast(new WebSocketServerCompressionHandler());
            pipeline.addLast(new WebSocketServerProtocolHandler(
                "/streaming", null, true));
            pipeline.addLast(new MeetupTopNSocketServerHandler());

            }
        });
        Channel channel = bootstrap
            .bind(bindAddress, PORT)
            .sync()
            .channel();

        System.out.println("Open your web browser to http://" +
            bindAddress + ":" + PORT + '/');

        channel.closeFuture().sync();
    }finally {
        bossGroup.shutdownGracefully();
        workerGroup.shutdownGracefully();
    }
  }
}
```

The port this will listen on

The IP address to bind to

❶ The event loops for Netty

The pipeline that will handle the request ❷

Our WebSocket ❸ handler

Bind the socket

Let the user know

Wait for things to close down

Shut down Netty when we're finished

It may seem like there is a lot going on in that code, but most of it is boilerplate Netty code. ❶ declares the event groups used by Netty, and ❷ does the basic configuration of them. ❸ is where we add `MeetupTopNSocketServerHandler` to the end of the pipeline. At a high level, Netty works by executing a pipeline that you construct, so we want to have our handler at the end of that.

`MeetupTopNSocketServerHandler` is the next class we need to look at. As you will see in the code download for this chapter, there is a lot of boilerplate code because it's dealing at a lower level with the networking Netty exposes. For that reason, we're only going to look at the relevant sections here. The first piece of code we want to look at is the handleWebSocketRequest method.

Listing 9.32 `handleWebSocketRequest` method

```
class MeetupTopNSocketServerHandler extends
        SimpleChannelInboundHandler<Object> {
                                                                    ❶ Map of
    ConcurrentHashMap<ChannelHandlerContext,StreamMessageConsumer>     context to
        channelToConsumer = new ConcurrentHashMap<>();                 consumer

    private void handleWebSocketRequest(ChannelHandlerContext ctx,
        WebSocketFrame frame) {
        //Ignoring the input.....
        //You would access it this way:
        //final String jsonRequest = ((TextWebSocketFrame) frame).text();

        if (!channelToConsumer.containsKey(ctx)) {                  Determine if
          StreamMessageConsumer streamMessageConsumer = new        client has already
            StreamMessageConsumer("meetup-topn-rsvps",             ❸ connected
                    UUID.randomUUID().toString(), ctx);
          channelToConsumer.put(ctx, streamMessageConsumer);
          streamMessageConsumer.process();                         Start to
        }                                                          ❹ process data

        TextWebSocketFrame returnframe = new
            TextWebSocketFrame(mapper.
                writeValueAsString("{response:success}"));         Return
          ctx.channel().write(returnframe);                        ❺ success
        }
}
```

Example of accessing input ❷ (points to line `//final String jsonRequest = ((TextWebSocketFrame) frame).text();`)

❶ is a map that we will use to tie together the connected channel with what turns out to be a Kafka consumer. If you wanted to have the client pass in information to your handler when they connect, ❷ is how you would go about accessing that passed-in data. ❸ is where the interesting part of this method happens—here we're checking to see if we already have seen this channel, and if not, we start up a StreamMessage-Consumer, which (as we will see shortly) results in a Kafka consumer being created. After constructing and stashing it in our cache, we need to tell it to start processing, as is done in ❹. When this is all set, in ❺ we return success to the client.

Now let's look at the relevant parts of the StreamMessageConsumer class, which is the workhorse of this service. The two most important parts are the process and run methods, both of them are shown in the following listing.

Listing 9.33 `process` and `run` methods

```
class StreamMessageConsumer {

    void process() throws Exception {                     Called by the
        messageProcessingThread.start();                  MeetupTopNSocket-
    }                                                      ❶ ServerHandler
```

```
      final class MessageProcessor extends Thread {
        @Override
        public synchronized void run(){
          while(!done){
            ConsumerRecords<String, String> records = consumer.poll(100);

            for(ConsumerRecord<String, String> record : records) {
              if (!channelHandlerContext.channel().isOpen()) {
                close();
              }else if(channelHandlerContext.channel().isWritable()) {
                channelHandlerContext.channel().writeAndFlush(new
                    TextWebSocketFrame(record.value()));
              }
            }
          }
        }
      }
    }
```

Try to consume from Kafka **2**

If the channel is closed, then clean up **3**

Write out the message to the client **4**

1 is called by the `MeetupTopNSocketServerHandler handleWebSocketRequest` that we looked at in listing 9.31. This will start an internal thread that processes the messages from Kafka. **2** is where we poll Kafka for new messages and then iterate over the records. If the WebSocket client that was connected to the service shuts down, we will catch that in **3** and in turn close down our Kafka consumer. Otherwise in **4** we will write the message received from Kafka to the socket.

We now have a streaming data API ready to go. To build it and run it, perform the next listing's commands in a new terminal window.

Listing 9.34 Building and running the streaming-data service

```
$> cd $EXAMPLE_CODE_HOME/Chapter9/streaming-api
$> mvn clean package
$> java -jar target/streaming-api-0.0.1.jar
```

You will now have the streaming API running. The next thing to do is start the collection service and deploy the `TopMeetupTopicsTopology`. Once those are up and running, the last thing to do is to bring up the UI and see the data flowing. To do that, open a browser and navigate to http://127.0.0.1:8080; you should see a UI that looks like figure 9.6.

See chapter 8 section 8.3.1

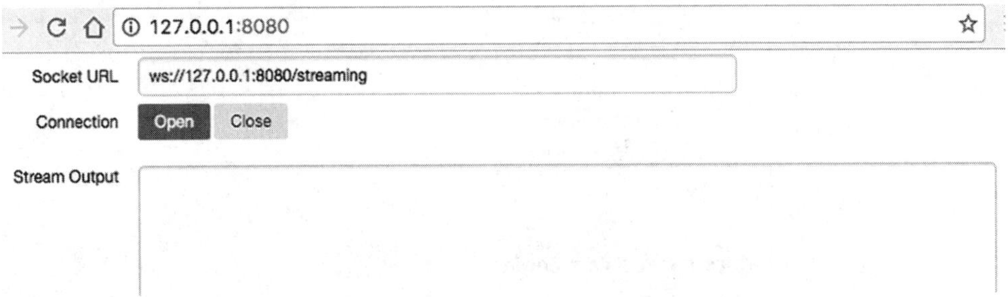

Figure 9.6 UI with no data

One more step to go. Click Open—if everything is working correctly, you should start to see data flowing, similar to what is shown in figure 9.7.

Congratulations! If you see data flowing like that in figure 9.7, you have a complete streaming-data pipeline running!

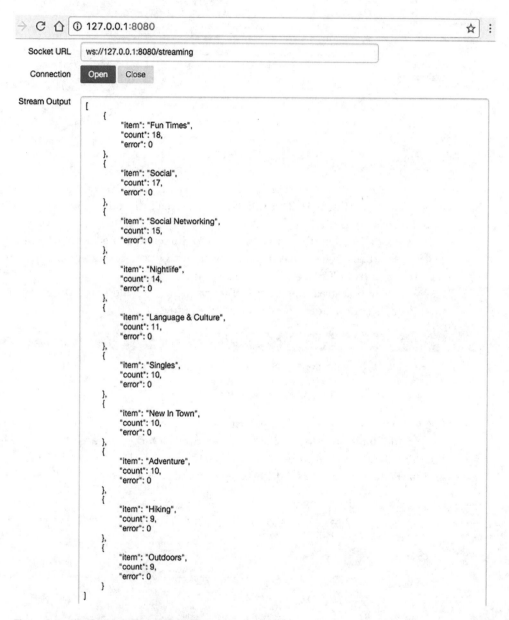

Figure 9.7 UI with streaming data flowing

9.5.1 *Taking it to production*

We do have a full streaming pipeline running and we built the streaming API with relatively little code. The following are some things you will want to think through and add to the streaming API before using it in a production deployment:

- There will be times when the client WebSocket isn't connected to the streaming API or fails to process the data you send. To overcome the loss of data, consider implementing the HML that we used in the collection service.
- Once many different clients connect to your service, you will want to have better control over the groups that are used when connecting to Kafka. In the example code for this section, we're using a 1:1 mapping between the WebSocket client and Kafka group. Instead of that, you will want to have a n:1 relationship—that way you can have many WebSocket clients consuming the same data from Kafka.
- Your users will want to start to query the data coming out of Kafka; you will want to look at adding that in. One way to do that is to allow the client to pass in a query when it connects and then save and apply that to data as it's flowing.

9.6 *Summary*

In this chapter we went from start to finish building out a streaming-data pipeline, taking note along the way of things you will need to consider as you build a production pipeline. With the material you've learned in the previous chapters along with this one, you should be well positioned to start building a streaming-data pipeline.

We learned the following:

- How to implement each of the tiers
- How to integrate each of the tiers
- Production considerations for various tiers

I hope you have enjoyed this chapter as much as I have enjoyed working on it; it is a lot of fun to see things come to life. Now is a good time to get a refill of your coffee and watch the stream flow in the UI.

index

MORE TITLES FROM MANNING

RxJS in Action
by Paul P. Daniels and Luis Atencio

ISBN: 9781617293412
400 pages
$49.99
May 2017

Rx.NET in Action
by Tamir Dresher

ISBN: 9781617293061
344 pages
$49.99
April 2017

Functional Programming in JavaScript
How to improve your JavaScript programs using
functional techniques

by Luis Atencio

ISBN: 9781617292828
272 pages
$44.99
June 2016

For ordering information go to www.manning.com

Reactive Design Patterns
by Roland Kuhn
   with Brian Hanafee and Jamie Allen

   ISBN: 9781617291807
   392 pages
   $49.99
   February 2017

Akka in Action
by Raymond Roestenburg, Rob Bakker,
   and Rob Williams

   ISBN: 9781617291012
   448 pages
   $49.99
   September 2016

Reactive Web Applications
Covers Play, Akka, and Reactive Streams
by Manuel Bernhardt

   ISBN: 9781633430099
   328 pages
   $44.99
   June 2016

For ordering information go to www.manning.com

MORE TITLES FROM MANNING

Spark in Action
by Petar Zečević and Marko Bonaći

> ISBN: 9781617292606
> 472 pages
> $49.99
> November 2016

Functional Programming in Scala
by Paul Chiusano and Rúnar Bjarnason

> ISBN: 9781617290657
> 320 pages
> $44.99
> September 2014

Functional Programming in Java
How functional techniques improve your
Java programs
by Pierre-Yves Saumont

> ISBN: 9781617292736
> 472 pages
> $49.99
> January 2017

For ordering information go to www.manning.com